Interpretation in Qualitative Research

Interpretation in Qualitative Research: Key Concepts in Qualitative Research is an edited collection that makes a methodological contribution to the conceptual connections around/with/for/to interpretation, and it offers contemporary concept provocations to the literature on interpretation.

Written for advanced undergraduate and early postgraduate students, as well as researchers seeking an entry point into a new area of study, it provides an introductory understanding of the key concept of interpretation in qualitative and postqualitative research across disciplines while working both with and against diverse interpretative traditions. This book touches upon different roles, functions, and reconfigurations of interpretation in (post)qualitative research and in different transdisciplinary, theoretical, and geopolitical contexts. Short, creatively crafted entries connect interpretation to its historical, discursive, and methodological roots while revisioning and reconfiguring what futures and becomings of interpretation might be like.

Interpretation in Qualitative Research can be used as a main text for introduction to qualitative research courses and advanced courses focusing on interpretation or interpretive sciences. It could also be used as a supplementary textbook for qualitative and postqualitative method courses, including data collection and research design courses.

Karin Murris is Professor of Early Childhood Education at the University of Oulu, Finland, and Professor Emerita of Pedagogy and Philosophy at the University of Cape Town, South Africa.

Mirka Koro is Professor of Qualitative Research at the Mary Lou Fulton Teachers College, Arizona State University, USA.

Key Concepts in Qualitative Methods

Series Editors: James M. Salvo
University of Illinois at Urbana-Champaign and Wayne State University, USA

Johanna Creswell Báez
University of Colorado at Colorado Springs, USA

Key Concepts in Qualitative Research is a dynamic book series designed to illuminate the essential theories, methodologies, and practices in qualitative research. Each volume provides a concise yet comprehensive exploration of a key concept, making it accessible for both novice and experienced researchers. With contributions from leading scholars, this series aims to bridge the gap between theory and application, fostering a deeper understanding of qualitative inquiry in various fields. The series is expertly edited by James Salvo and Jasmine Ulmer, whose commitment to advancing qualitative research ensures a rich and engaging reading experience. Proposals and manuscripts should be directed to salvo3000@gmail.com or jbaez2@uccs.edu

Issues of Equity
Key Concepts in Qualitative Methods
Johanna Creswell Báez

Critical Qualitative Research & Social Justice
Key Concepts in Qualitative Methods
Penny A. Pasque

Interpretation in Qualitative Research
Key Concepts in Qualitative Methods
Edited by Karin Murris and Mirka Koro

For more information about this series, please visit: https://www.routledge.com/Key-Concepts-in-Qualitative-Methods/book-series/KCIQM

Interpretation in Qualitative Research

Key Concepts in Qualitative Methods

Edited by Karin Murris and Mirka Koro

Routledge
Taylor & Francis Group
LONDON AND NEW YORK

Designed cover image: Thibaud LEROUX / Getty images

First published 2025
by Routledge
4 Park Square, Milton Park, Abingdon, Oxon OX14 4RN

and by Routledge
605 Third Avenue, New York, NY 10158

Routledge is an imprint of the Taylor & Francis Group, an informa business

British Library Cataloguing-in-Publication Data
A catalogue record for this book is available from the British Library

Library of Congress Cataloging-in-Publication Data
Names: Murris, Karin editor | Koro-Ljungberg, Mirka editor
Title: Interpretation in qualitative research : key concepts in qualitative methods / edited by Karin Murris and Mirka Koro.
Description: Abingdon, Oxon ; New York, NY : Routledge, 2026. | Series: Key concepts in qualitative methods | Includes bibliographical references and index. | Summary: "Interpretation in Qualitative Research: Key Concepts in Qualitative Research is an edited collection that makes a methodological contribution to the conceptual connections around/with/for/to interpretation and it offers contemporary concept provocations to the literature on interpretation.It can be used as a main text for introduction to qualitative research courses, advanced courses focusing on interpretation or interpretive sciences. It could also be used as a supplementary textbook for qualitative methods courses including data collection and research design courses"— Provided by publisher.
Identifiers: LCCN 2025012222 (print) | LCCN 2025012223 (ebook) | ISBN 9781032902715 hardback | ISBN 9781032954622 paperback | ISBN 9781003585046 ebook
Subjects: LCSH: Qualitative research | Qualitative research—Methodology
Classification: LCC H62 .I658 2026 (print) | LCC H62 (ebook)
LC record available at https://lccn.loc.gov/2025012222
LC ebook record available at https://lccn.loc.gov/2025012223

ISBN: 978-1-032-90271-5 (hbk)
ISBN: 978-1-032-95462-2 (pbk)
ISBN: 978-1-003-58504-6 (ebk)

DOI: 10.4324/9781003585046

Typeset in Optima
by Apex CoVantage, LLC

To life and relationality especially outside the Academy, to water and snow.

Contents

Figures

Acknowledgments

We would like to thank James Salvo and Jasmine Ulmer for the invitation to co-edit this book on interpretation in the Key Concepts series. As editors, we would like to thank in particular our section editors Viv Bozalek, Kerryn Dixon, Karen Malone, Roger Norum, and Jayne Osgood for their help in sourcing authors. Originally conceptualized as part of an online encyclopedia, the contributions to this book have been through many iterations, taking several years to complete. Moreover, out of the Key Concepts project grew the idea to publish a more in-depth, theoretically elaborated, and expansive book by the same editors under the title *Handbook of Reconfiguring Interpretation in Postqualitative Research*. This gave the authors the opportunity to explore their content and theoretical connection at a deeper level and further build the ideas they touched upon in the entries covered in this more introductory *Key Concepts* book. Furthermore, the expanded Handbook also enabled the inclusion of other authors and perspectives.

We would also like to thank the contributing authors for their continuing commitment to this project and their gracious patience when the COVID-19 pandemic severely delayed its completion. Thank you, friends, and friends who care for you! Without the human and other-than-human relations we and they are part of, this book would not have been possible. We are delighted with the final entries and how the alphabetically organized 'subconcepts' offer provocative opportunities to explore interpretation – one of the key concepts in research. The entries 'crack the concept open' in often creative, sometimes provocative ways, reconceptualizing and rethinking interpretation as a situated practice of multiplicities.

The editors, section editors, and authors also deserve credit for acting as reviewers for the chapters of their peers. All chapters were double-blinded peer-reviewed with the extra assistance of other external reviewers, for which we are very grateful.

Thanks also to the editors of Routledge, Eleanor Taylor, Adam Woods, and Matt Bickerton, for their continuous assistance, support, and guidance.

Contributors

Annette Arlander, DA, is an artist, a researcher, and a pedagogue. She is a former professor of performance art and theory and professor of artistic research at the University of the Arts Helsinki and former professor of performance, art, and theory at the Stockholm University of the Arts. Her research interests include artistic research, performance as research, and the environment. Her artwork moves between the traditions of performance art, video art, and environmental art. See https://annettearlander.com.

Teresa K. Aslanian is Professor of Early Childhood Education in the Department of Educational Science, University of South-Eastern Norway. A trained preschool teacher with 15 years of practice in the field, Aslanian has published widely, engaging with various philosophical perspectives to explore and reconceptualize central themes in early childhood education and care (ECEC), such as children's engagements with love, care, play, and holistic learning. She is committed to challenging the ideas of ECEC as a necessary site for children's growth and development. She has edited special issues and several anthologies, including *Children and the Power of Stories: Posthuman and Autoethnographic Perspectives in Early Childhood Education* (2022). She serves on the editorial board of *Reconceptualising Education Research Methodology*.

Randall Bird (PhD) has worked on the relationship between African art, architecture, and landscape and the dynamics of colonialism, with a focus on Madagascar. For ten years, Bird lived in South Africa, where he taught the history of architecture in the professional MArch degree programs at the University of the Witwatersrand and the University of Johannesburg. He serves as editor of the Africa volume for the *Encyclopedia of Vernacular Architecture of the World* (Bloomsbury Press, forthcoming 2026). Currently, he is engaged in archival research for a study devoted to the seminal role that the founding director of the New York Public Library, John Shaw Billings, played in the architectural design of the central building. Bird completed his MArch degree at Washington University in St Louis, an MA in the history of science, and a PhD in the history of art and architecture at Harvard University.

Ruth Boycott-Garnett (PhD) works with the spatial and material encounters of babies and toddlers. Her research stems from 15 years of working with performance and story-making across multi-professional networks to develop creative practice for 0–3s. Her doctoral research focused on babies as space makers and how attending to babies' spatial practices allows a rethinking of public space. She is particularly interested in how paying attention to babies' and toddlers' daily lives brings about methodological invention and how visual and arts-based methods can work generatively with babies and toddlers in research.

Vivienne Bozalek is Professor Emerita in Women's and Gender Studies at the University of the Western Cape and Honorary Professor in the Centre for Higher Education Research, Teaching and Learning (CHERTL) at Rhodes University. Her most recent coauthored book with Michalinos Zembylas is *Responsibility, Privileged Irresponsibility, and Response-ability: Higher Education, Coloniality and Ecological Damage* (Palgrave, 2023), and her most recent co-edited book is *Hydrofeminist Thinking with Oceans: Political and Scholarly Possibilities* with Tamara Shefer and Nike Romana (Routledge, 2024).

Svend Brinkmann is Professor of Psychology at the University of Aalborg, Denmark, and works on the philosophy and methodology of psychology and related disciplines.

Abhik Chakraborty (Ph.D.) is currently an associate professor at the Faculty of Tourism, Wakayama University, Japan. His research interests include mountains, geological time, the Anthropocene, global environmental change, UNESCO WNHS, and tourism and environmental change in natural heritage landscapes. The key focus of his research is on generating a qualitative understanding of environmental/Earth system change. He has written several papers on these issues and was the lead editor for the coauthored volume *Natural Heritage of Japan* (Springer, 2018). His other notable passions are mountaineering, trail running, and landscape photography. (From: https://www.tandfonline.com/doi/full/10.1080/1743873X.2023.2169614.)

Julianne Cheek is Professor Emerita at Østfold University College, Norway. She is Editor-in-Chief of *Qualitative Health Research* and an international advisory board member and editorial board member of a number of books and journals related to qualitative inquiry. Her publications reflect her ongoing interest in qualitative inquiry and the politics of that inquiry.

Ruth Churchill Dower has recently completed her Ph.D. scholarship at Manchester Metropolitan University (UK), exploring young children's nonlingual ways of being through experiments in movement. Ruth is interested in the sensing practices, pedagogies, and relations that facilitate young children's more-than-human connections. Her book *Creativity and the Arts in Early Childhood* explores the origins, impacts, and conditions for creative potential to thrive and how different art forms can be nurtured in early childhood practices. Ruth is the founder of Earlyarts, an online training hub for early childhood and arts professionals, as well as a musician, storyteller, mountain lover, and former teacher.

Sarah Crinall (she/her) is a philosopher in education and children's advocate for more messy, situated, and relatable initial teacher education, education research methodologies, and sustainable education policy futures. Sarah is a place-based iphonographer, a marine ecologist, and a citizen scientist who works with art as muse. She lives on Millowl by Warn Marin, Boonwurrung Country, Kulin Nation, which are the unceded lands of Yallock-Bulluk peoples, Australia, with her two daughters and partner, swamp paperbarks, and black swans. Sarah is part of an Anglo-Indian/Australian family who have lived and moved across multiple colonies and coastlines on the south-eastern coast of so-called Australia.

Barbara Dennis (she/her) is a feminist critical scholar of qualitative inquiry particularly concerned with justice, onto-epistemologies, and ethics. She is a grandmother and a professor. She is a seamstress and a beekeeper and is a lover of stories. She has been professionally situated at Indiana University.

Alycia Elfreich (she/her) is Assistant Professor of Multicultural Education at Montana State University. She draws from critical feminist and relational qualitative methodologies to examine justice-oriented curriculum and pedagogy in teacher education, curriculum studies, and school-community contexts. Her professional and personal commitments lie in spaces where building relationships and meaningful community honor the complexity of the human experience and opportunities to listen and learn with one another.

Darcy Furlong (she/her) is a Ph.D. candidate in the Qualitative Methodology program in the School of Education at Indiana University, Bloomington. She draws upon disability theory to inform her methodological understandings and anti-ableist research practices. In her applied work, she uses conversation analysis to study healthcare interactions, particularly within neurodisability contexts.

Michael Gallagher is Reader in the Faculty of Health and Education at Manchester Metropolitan University. His research is interdisciplinary, exploring sonic methods, sonic geographies, children and young people's lives, and theory and methodology in qualitative research. He has published widely across journals in human geography, childhood studies, sound studies, education, allergy, social work, and music. Outside of his academic work, Michael produces electronic music, provides audio mastering services, and writes about music technologies.

Susanne Gannon discovered collective biography during her doctoral research (with Professor Brownyn Davies) on feminist post-structural modes of writing research. She has taken up collective biography as a capacious, embodied, and creative method for working with memories and exploring subjectivities and has continued to work with and develop the method with collectives in Germany, England, Canada, Australia, Finland, Sweden, Denmark, and more over several decades. Susanne is Professor of Education at Western Sydney University, Australia. Her current research encompasses a range of equity and social justice aspects of secondary schooling.

Abigail Hackett is Professor of Childhood and Education at the Sheffield Institute of Education, Sheffield Hallam University. She is interested in the role of place, materiality, and bodies in young children's lives. She researches mostly in community spaces, in collaboration with children and families, employing ethnographic and post-qualitative methods. She has researched early years pedagogy in Norway, children's experiences of trees in South Yorkshire, and young children's literacy practices as part of an international project bridging the United Kingdom, Finland, and Australia. She is the lead editor of *Working with Young Children in Museums: Weaving Theory and Practice*, published with Routledge, and author of *More-than-Human Literacies in Early Childhood*, published with Bloomsbury.

Joanna Haynes (PhD) is Associate Professor of Education Studies at Plymouth University Institute of Education. Her background is in philosophy, and her research interests are in democratic and community education, age and intra-generational relations, children's literature, and philosophy with children. Her book *Children as Philosophers* (2002) has been translated into Greek, Spanish, and Korean; *Picturebooks, Pedagogy and Philosophy* (2012) has been translated into Arabic; and *Literacies, Literature and Learning: Reading Classrooms Differently* (2018) was coedited with Karin Murris. She coedited *The Routledge International Handbook of Philosophy for Children* (2017). Current research projects she is leading include the *Between*

Ages Project (https://www.e-publicacoes.uerj.br/childhood/article/view/86815) and the BERA-funded *Valuing Disagreement. Teacher Education and Democracy* project. Together with Jocey Quinn, Joanna coordinates the Adventures in Posthumanism network. https://www.plymouth.ac.uk/research/education/adventures-in-posthumanism.

Shakhnoza Kayumova is Associate Professor of STEM Education and Teacher Development and Director of the Kaput Center for Research and Innovation in STEM Education at the University of Massachusetts Dartmouth. Her research focuses on the exploration of conditions of possibility for developing asset-based identity spaces in STEM education and the role of diverse onto-epistemological and methodological approaches in fostering expansive and transformative teaching and learning spaces.

Walter Omar Kohan is Full Professor of Philosophy of Education at the State University of Rio de Janeiro (Brazil); Senior Research Member of CNPQ, FAPERJ, and CAPES, Brazilian Ministry of Education; and Former President of the International Council for Philosophical Inquiry with Children (ICPIC). Since 2005, he has been the coeditor of *Childhood & Philosophy* (http://www.e-publicacoes.uerj.br/index.php/childhood). Furthermore, since 2002, he has been organizing the Biennial International Colloquiums of Philosophy and Education in Rio (last: *filoeduc.org/12cife*). Some of his main publications in English include *Philosophy and Childhood: Critical Perspectives and Affirmative Practices* (Palgrave, 2014); *Childhood, Education and Philosophy: New Ideas for an Old Relationship* (Routledge, 2015); *The Inventive Schoolmaster* (Sense, 2015); *Thinking, Childhood, and Time: Contemporary Perspectives on the Politics of Education* (edited with Barbara Weber; Lexington, 2020); and *Paulo Freire: A Philosophical Biography* (Bloomsbury, 2021). Some of his recent publications in Portuguese include *Paulo Freire: um menino de 100 anos* (NEFI, 2021, free download at *filoeduc.org/editora*) and *A viagem dos sonhos impossíveis* (Autêntica, 2022).

Mirka Koro (Ph.D., University of Helsinki) is Professor of Qualitative Research at the Mary Lou Fulton Teachers College, Arizona State University. Her scholarship operates at the intersection of qualitative inquiry, methodologies, philosophy, experimentalism, and socio-cultural critique. She has published in various qualitative, methodological, and educational journals. She is also the author of *Reconceptualizing Qualitative Research: Methodologies without Methodology* (2016) and *Knowledge Production in Material Spaces: Disturbing Conferences and Composing Events* (2022) and coeditor of *Disrupting Data in Qualitative Inquiry: Entanglements with the Post-Critical and*

Post-Anthropocentric (2017) and *Intra-Public Intellectualism: Critical Qualitative Inquiry in the Academy* (2021).

Teija Löytönen holds a Doctor of Arts from the University of Arts Helsinki, Theatre Academy in Finland. She has been a funded scholar (Research Council of Finland) at both the University of the Arts Helsinki and Aalto University, as well as a Fulbright scholar at Arizona State University, USA. Her research interests include higher arts education, arts, and creativity in academia, as well as professional and academic development. Her special interest is in collaborative research endeavors and the diversity of qualitative research. She has published in several international refereed journals and edited volumes as well as presented her research in various academic networks.

Christina MacRae (PhD) is Research Fellow at Manchester Metropolitan University. Her research interests lie in early childhood with a focus on sense, affect, and movement in children's world-making practices. She has also been involved in research that has focused on young children's learning in classroom and nursery contexts, as well as in museums, galleries, and outdoor public spaces.

Karen Malone (PhD) is Professor of Education and Environmental Philosophy at Swinburne University. As an international author, educator, and researcher, she engages in studies of the ecological crisis, focusing on human-environment relations over time and the shifting view of more-than-human agency. She is interested in understanding the ecological crisis, particularly the impacts of climate change, human-initiated disasters, and biodiversity loss on human and non-human lives in places. In her teaching and research, she applies ecofeminist, post-human, vital materialist, and Indigenous theoretical perspectives, using post-qualitative encounters and earth-focused pedagogical practices of walking with and paying attention. Her most recent books include *Wild Ecologies: Walking-with Glaciers, Theorising Posthuman Childhood Studies,* and *Children in the Anthropocene*.

Karin Murris (PhD) is Professor of Early Childhood Education at the University of Oulu (Finland) and Professor Emerita of Pedagogy and Philosophy at the University of Cape Town (South Africa). Grounded in academic philosophy, her main research interests are in philosophy of education, posthumanist child, and death studies. Karin has extensive experience in undertaking a wide range of funded research by national governments, charities, and industry. Karin has (co-)authored the following books: *Postdigital Play and Global Education: Reconfiguring Research* (2024); *Karen Barad as Educator: Agential Realism and Education* (2022); *The Posthuman Child* (2016); *Literacies,*

Literature and Learning: Reading Classrooms Differently (2018); and *Picturebooks, Pedagogy and Philosophy* (2012). Furthermore, she has (co-)edited *Navigating the Postqualitative, New Materialist and Critical Posthumanist Terrain across Disciplines* (2021); *Glossary: Postqualitative: New Materialist and Critical Posthumanist Research across Disciplines* (2022); and the *Routledge International Handbook of Philosophy for Children* (2017). She is Chief Editor of the Routledge Postqualitative, New Materialist and Critical Posthumanist Research series (https://postqualitativeresearch.com).

Denise Newfield has been a teacher educator and researcher in the field of literature and literacy education all her life – for over 40 years as a lecturer and professor at the University of the Witwatersrand, South Africa. Her Ph.D. (Institute of Education, University of London) is titled *Transmodal Semiosis in Classrooms.* Her official retirement has been filled with outreach and research projects such as ZAPP (the South African Poetry Project), Phenduka (Alexandra Township Literacy Project), and Doing Academia Differently (with Vivienne Bozalek). Always interested in change and transformation rather than systemic theory and fixed method, and in this way challenging static, colonial educational practices, her work in recent years has been fueled and developed by encounters with post-philosophies and new feminist materialist theories. Her work has thus moved away from systematic analytical approaches to more fluid, experimental, and process-orientated approaches, as evidenced in the chapter in this book, 'Multimodalities' (co-written with Sarah Crinall).

Pilvi Lumia Porkola, DA, is an artist, a researcher, a pedagogue, and a writer. Currently, she works as University Researcher at Uniarts Helsinki. The current topic of her research is Performance Pedagogy – Politics of Art Pedagogy. Her research interests include performance art, performance studies, artistic research, arts-based methods, and intersections between arts and sociology. As an artist, she is interested in everyday life, performing, and objects.

Pauliina Rautio is Professor of Biodiversity Education at the University of Oulu. She currently leads several large-scale projects on multispecies everyday lives in the framework of environmental social sciences. Her transdisciplinary research team AniMate includes scholars across social and natural sciences, also collaborating with artists and producing in-depth theoretical-empirical studies of becoming and being human with other animals.

Rose-Anne Reynolds has a Ph.D. in education from the University of Cape Town (UCT). She is an early childhood education lecturer in the

School of Education at the UCT. Rose-Anne's transdisciplinary Ph.D. thesis is titled 'A Posthuman Reconfiguring of Philosophy with Children in a Government Primary School in South Africa'. Her research interests include philosophy with children (P4wC), the philosophy of child and childhood, pedagogies of enquiry, early childhood education, post-qualitative research, and inclusive education, including disability studies. Some of Rose-Anne's publications can be found here at http://www.education.uct.ac.za/rose-anne-reynolds.

Tamara Shefer is Professor of Women's and Gender Studies in the Faculty of Arts and Humanities at the University of the Western Cape, Cape Town. Her scholarship has foregrounded the study of gender and sexualities within postcolonial, decolonial, and transnational feminist and critical masculinities thinking, with particular emphasis on young people. She is also engaged with reconceptualizing academic knowledge, with an emphasis on embodied, affective, feminist, and decolonial pedagogies and research, including engaging ethics of care in social and environmental justice. Her current work foregrounds collaborations across art and activism and thinking with oceans and water. Her recent edited volumes include *Hydrofeminist Thinking with Ocean/s: Political and Scholarly Possibilities* (Shefer, Bozalek, and Romano, Routledge, 2024) and the *Routledge Handbook of Global Feminisms and Gender Studies: Convergences, Divergences and Pluralities* (Torres, Pinto, Shefer, and Hearn, 2024). Her recent monographs include *A Feminist Critique of Sexuality Education for Gender Justice in South African Contexts* (Shefer and Ngabaza, 2023, CSA&G Pretoria University) and *Knowledge, Power and Young Sexualities: A Transnational Feminist Engagement* (Shefer and Hearn, 2022, Routledge).

Kathryn Strom (Katie) is Associate Professor of Educational Leadership at California State University, East Bay, and Director of CSUEB's Center for Research on Equity and Collaborative Engagement (CRECE). Her research combines multiple critical and complex theories to study teacher learning practice and to advocate more broadly for more relational, difference-affirmative ways of thinking-being-doing in education and academia.

Juha Suoranta is Professor of Adult Education at Tampere University, Finland. He studies and teaches critical pedagogy and public sociology. His books include *Hidden in Plain Sight: How I Sheltered a Refugee* (2011), *Artistic Research Methodology* (2014), *Militant Freire* (2021), *Paulo Freire, Peregrino do Óbvio* (2025), Finnish *Rebellious Research* (2014 in Finnish), and *C. Wright Mills's Sociological Life* (2017 in Finnish).

Tuure Tammi is a post-doctoral researcher at the University of Oulu and an adjunct professor of education at the University of Tampere, Finland. Drawing from relational ontology and critical post-humanist and multispecies scholarship, his work explores the theoretical, methodological, and philosophical questions arising from the specific circumstances where species meet. He is a co-leader of the transdisciplinary research team AniMate, editor for *Trace – Journal for Human-Animal Studies*, and a sonic artist working for independent record labels.

Ananí M. Vasquez, Ph.D. (Arizona State University), combines her experiences in general, bilingual, gifted, and special education(s) to envision an inclusive education. She draws on creativity theory, disability studies in education, the neurodiversity paradigm, process philosophy, and arts-based inquiry while working with others toward post-oppositional educational transformation. Ananí has developed creative professional learning experiences based on 'micromoments' (surprising, unexpected instances in learning events) that encourage teachers to reflect on ableist structures, routines, and practices in schooling and to envision possibilities for shifting classroom interactions in their own contexts. Ananí is also a school site instructional coach, facilitating professional learning teams (PLTs) and coaching K-8 teachers. Her publications include articles and chapters on qualitative research and disability studies in education, and she is the coeditor of *Writing and the Articulation of Postqualitative Research* (2023).

Henrika Ylirisku (DA, Doctor of Arts) is an art pedagogue and researcher-artist, serving as a university lecturer of art education at Aalto University in Finland. Her work lies at the intersection of arts and art education, environmental education, and multispecies research. Recently, Henrika has published on the theoretical groundings of environmental art education and multispecies forest relations. Her current research explores shifting nature-culture relations and the atmospheres of young people growing up amidst the environmental crisis. She also engages in movement-based practices and has developed artistic walking methodologies, inspired by the working methods of environmental and performance arts.

Outi Ylitapio-Mäntylä's research focuses on gender equality in educational communities, and her research interests include power and agency, particularly in early childhood education. Her methodological interests include narrative approaches and the method of memory work, which she has also applied as a teaching method. Her recent research has focused on memories of girlhood and peer mentoring of educational professionals. Outi (Ph.D., adjunct professor) works as a university lecturer in the Faculty of Education and Psychology at the University of Oulu.

Michalinos Zembylas is Professor of Educational Theory and Curriculum Studies at the Open University of Cyprus and Honorary Professor at Nelson Mandela University, South Africa. He holds a Commonwealth of Learning (COL) Chair for 2023–2026. He has written extensively on emotion and affect in education, particularly in relation to social justice, decolonization, and politics. His latest books are *Responsibility, Privileged Irresponsibility and Response-ability in Contemporary Times: Higher Education, Coloniality and Ecological Damage* (co-authored with Vivienne Bozalek) and *Working with Theories of Refusal and Decolonization in Higher Education* (coedited with Petra Mikulan).

Mnemo ZIN is a composite name for Zsuzsa Millei (Tampere University), Iveta Silova (Arizona State University), and Nelli Piattoeva (Tampere University). By adopting a collective name, we foreground our entangled, perpetual becoming-with as researchers and human beings who refuse to single out or rank our contributions. Our collective name is inspired by the figure of Mnemosyne from Greek mythology, goddess of memory and mother of the nine Muses. Spanning almost ten years, our research examines childhood memories through the collective biography method, writing alternative histories, and informing our current thinking about (post)socialist and (de)colonial pasts, presents, and futures.

Introduction

Situating Interpretation in Conceptual Spaces

Karin Murris and Mirka Koro

Interpretation could be seen as one of the main concepts of "traditional" qualitative inquiry, ethnographic processes, and anthropological research approaches. Our purpose in this book is to conceptualize and engage with interpretation as a practice of multiplicities. More specifically, this book addresses different roles, functions, and reconfigurations of interpretation in (post)qualitative research also as envisioned in the future(s). Diverse and creatively crafted entries connect interpretation to its historical, discursive, and methodological roots while revisioning and reconfiguring what futures and becomings of interpretation might be like in different transdisciplinary, theoretical, and geopolitical contexts.

Deleuze and Guattari (1994) noted that there are no simple concepts; concepts have multiplicity and relationality, and concepts function as fragmentary wholes. Similarly, this book does not seek to offer *one* conceptual understanding of interpretation. Rather, it offers multiple and diversifying insights, examples, and practices of how interpretation could be practised and lived within and across chapters and in different conceptual spaces. Overall, concepts of and associated with interpretation "link up with each other, support one another, coordinate their contours, articulate their respective problems, and belong to the same philosophy, even if they have different histories" (Deleuze & Guattari, 1994, p. 18). Concepts are also situational and speak to the event without references and discourses. They only offer resonations.

Similar to Deleuze and Guattari, who view concepts as the archipelago, skeleton frame, formless surfaces, and concrete assemblages, interpretation conceptualizations and practices take different forms. Interpretation concepts are in continuous flux, and thus, they "pave, occupy, or populate" (Deleuze & Guattari, 1994, p. 36). They create.

Furthermore, concepts in this book do not function as universal but pluriversal. Interpretation as a pluriversal concept does not have definitions, scholarship, authority, discourse, or geographical location but several authors, locations, discourses, and practices. The plurality of interpretation, different theoretical contexts shaping interpretative

DOI: 10.4324/9781003585046-1

practices, and other transformations across time have influenced the role and function of interpretation in qualitative and postqualitative inquiry. Following Manning (2020), concepts in this book cannot be separated from the thoughts, acts, and processes they provoke. Concepts are not ready or made ahead of time. Instead, they result from the activation of ecologies interacting with them. Concepts of this kind form intensities, and to "find a concept is to touch on a nerve of experience, to catch the necessity of its naming" (p. 11). In this book, scholars have followed the concept of (qualitative) interpretation to move towards various places of affective difference through their carefully crafted subconcepts.

More specifically, interpretation can be seen as one of the cornerstones and historical markers of qualitative inquiry and research. Thus, today's uses/applications/extensions of interpretation look very different from their 1970s and 1980s counterparts. The earlier interpretation was heavily situated in phenomenologies, hermeneutics, and constructivisms, whereas today's interpretation has a more material and more-than-human focus. One could argue that many qualitative scholars still "interpret" or work through data/materials/lives, but the processes and aims of these practices have shifted. What happens to interpretation when put into different theoretical and philosophical frames, including postmodernism, posthumanism, and postcolonialism? What has changed, and how do different epistemological and ontological spaces offer different insights about interpretation and shape diverse uses for interpretation? How does interpretation function, and what does it produce during these rapidly shifting cultural and political times?

Postqualitative researchers often affirm and build on qualitative researchers' work (e.g., constructivist and phenomenological) who themselves also resist taking power-producing binaries for granted, such as subject/object, culture/nature, human/technology, or human/matter. Practices of interpretation do not take place in a vacuum but are part of existing relations, dynamically forming new relations and relational practices. Relationality, in turn, is shaped by the actively involved actors, subjects, matter, and diverse other forces.

The bifurcation between nature and culture is deeply entrenched within Western culture. It presupposes that thinking is about representing, measuring, calculating, mapping, and picturing a world that is given or simply "there" (Barad, 2007). The non-/other-/more-than-human is, by the very same token, positioned as an inert, passive substance and concept. However, our research practices are not politically innocent and orderly. Rather, the alphabetically organized subconcepts (of interpretation) in his book have been carefully chosen as activators assisting readers and users in interrogating constructively and affirmatively the key concept of interpretation. The idea is to engage with these diverse concepts, making possible connections to interpretation processes through our research practices, and not just theoretically. Concepts such as voice, criticality, diffraction, digitalization, multimodality, multispecies, and mountains generate fresh opportunities to trouble the anthropocentric nature of interpretation within our research practices that thrive and sediment even further the culture/nature binary. Other concepts such as architecture, collective biography, community of philosophical enquiry, deconstruction, (no) method, history, performance, posts, wriggling with data, languages, and politics help us become (uncomfortably) aware of how the research methods and approaches we tend to use in academia assume certain kinds of *a priori* relationalities. These relations are of an individualized and hierarchical nature, not only between word and world but also between question and answer and human and non-/other-/more-than-human entities. As illustrated in this book, interpretations are always positioned in relation to time-space-matterings. Therefore, situating interpretation in conceptual spaces has implications for analysis and how we engage as researchers with data. What *Interpretation in Qualitative Research* provokes us to do is not to reject interpretation (nor the interpretative turn) altogether but to reconfigure interpretation and to reimagine ways of doing research that bring about different future(s) – a commitment, Donna Haraway (2016) reminds us, to some worlds, rather than others.

References

Barad, K. (2007). *Meeting the Universe Halfway: Quantum Physics and the Entanglement of Matter and Meaning*. Duke University Press.

Deleuze, G., & Guattari, F. (1994). *What is Philosophy?* (H. Tomlinson & G. Burchell, Trans.). Columbia University Press.

Haraway, D. J. (2016). *Staying With the Trouble: Making Kin in the Chthulucene*. Duke University Press.

1 Activism

Tamara Shefer

Interpretation in/Through/With Activist-Knowledges

This entry argues the value of 'thinking' with overt forms of activism for scholarly interpretation. Activist and aesthetic activisms are explored as productive spaces of interpretation for (post)qualitative research. There is a large and growing body of feminist and other critical justice work that argues the value of collaborative thinking and organising between researchers with activists and artists and challenging these binaries of discipline and modality (e.g., Buikema, 2021; Lalu, 2023; Nagar, 2014, 2019; Pather & Boulle, 2019; Sangtin Writers Collective & Nagar, 2006; Truman, 2022). Here, I gesture to such collaborations across, between, and through art-activism, as both an ethical and rich space for interpretation that also challenges the binary of academic knowledge and justice activism that is still key to the dominant Cartesian logics of the university.

In 2015, South African-based artist-activist Sithembile Msezane performed *Chapungu – The day Rhodes Fell* (http://www.sethembile-msezane.com/kwasukasukela). This installation took place while the statue of Cecil John Rhodes, the most infamous colonial figure in South Africa, was removed from its central place at the University of Cape Town (UCT) after some weeks of protest by #RhodesMustFall activists calling for its removal (see also Buikema, 2021; Murris, 2016; Romano, 2019; Shefer, 2018). Msezane stood on a plinth near the Rhodes statue for many hours while the statue was being removed. She was dressed to represent the soapstone statues of the African bird that were removed from the ruins of Great Zimbabwe during the colonial period and sold to wealthy settlers, like Rhodes (Hubbard, 2009). Msezane's performance is an activist interruption that presents 'a powerful material-discursive disruptive moment in reminding the viewers of the multiple layers of colonial theft and violences' (Shefer, 2018, p. 174). The installation also gestures to a disruption of the imperialist civilising project that underpins the pursuit of mainstream academic knowledge. In this decolonial moment, the ascendancy of African femininity, the

DOI: 10.4324/9781003585046-2

toppling of white, male, northern settler authority, and a symbolic 'correction' of that which was violated are extended to a disruption of colonial and patriarchal power in the academy and elsewhere. As Karin Murris (2016, pp. 281–282) points out:

> Msezane's youth, race and gender stand in stark contrast with the older, white, male Cecil Rhodes. It is this juxtaposition that provokes a response, to take response-ability for the a/effects of power producing binaries and the marks they make on bodies.

Deploying the imagery of a stolen artwork that represents an animal is perhaps not only talking back to colonial abuses but also gestures to and exposes the intersectionalities of 'subalternity' (Spivak, 1988) – colonised 'subjects', femininity, other species (the Great Zimbabwe Bird represented in the statue), and non-sentient beings (a statue) – while also challenging the human exceptionalist, 'civilising project' driven by universities historically and in current global capitalist times. Murris's (2016) article also helpfully places a human-centric reading in juxtaposition with a post-humanist reading of activism related to the statue's removal, which speaks to the diverse ways in which activist art may work in interpretations depending on the philosophical approaches of the researcher.

This is but one example of the richness of 'thinking' with activism for scholarly interpretation, which is the provocation of this piece. Here, I make a case for the valuable work between and intra-action of interpretation and activism. How does interpretation guide and work within and through activism? How do we interpret, for whom and what, towards what ends, and in ways that do *not* reproduce the very violences we hope to challenge?

It is important to flag that while advocating a 'thinking' with activism for qualitative interpretation, what counts as activism is also contested. In challenging binaried thinking which separates knowledge from other forms of meaning-making, activism as separate from interpretation needs to be troubled in the first place. Deploying Barad's notion of agential realism renders all interpretations a form of activism since the search for meaning is already a search for justice.

With respect to the complex relationship between knowledge and justice, there has been a long, intensifying critique of the normative colonial and masculinist scholarly practices that are dominant in global norms of interpretation in scholarship. Troubling research and its ontologies, epistemologies, and methodologies of interpretation are also bound up with the project of challenging which knowledges count and which knowledges, where and by whom, are regarded as authoritative and 'proper' knowledge. Feminist, queer de-postcolonial scholarship has been involved in a long journey of destabilising the authoritative, expert scholar and working in collaborations across disciplines and modalities towards reconceptualising scholarship. New materialist, post-humanist, and post-qualitative scholarship has further strengthened calls for alternative ways of knowledge that resist the representational and extractivist logics that persist in dominant forms of interpretation.

Key to disrupting the 'troubling' dynamics of research and related practice is a deeper engagement with reconceptualising academic norms of research and interpretation of 'data'. This means going beyond the praxis of reflexivity (Lenz Taguchi, 2010) and empathy (Hemmings, 2012) that feminist and qualitative researchers have promoted for decades to a radical reimagination of what it means to do research and interpretation that does not rely on the objective subject and Cartesian divides. Central to this task are post-humanist, materialist, decolonial feminist thinking and also centuries of indigenous knowledges that engage a sensibility of human entanglements (and therefore response-ability) with each other and with other species, non-living entities, and planetary environmentalities. A further provocation of such alternative interpretative epistemologies is to disrupt the binaries that characterise dominant logics of interpretation, that is the Cartesian divide of mind-body, rationality-affect, and logic-creativity, and to particularly re-engage the productive and silenced negative components of such binary logic. In the dominating patriarchal logic of the academy, hugely intensified by neoliberal capitalist imperatives, bodies, affect, and relationalities are denigrated and excluded through the privileging of mind, rationality, and individualist 'outputs' and careers. In meeting these provocations, we suggest imagining and experimenting with ways to do scholarly work, including what has been known as 'interpretation' that challenges the epistemic violence of academic interpretation and finds 'ways in which this system can be interrupted through a self-conscious and deliberate set of pedagogies, methodologies, and tactics' (Nagar, 2013, p. 3).

One terrain for such an interruption, in deploying the opening example, is through the overt conversation between research and activism: indeed, to begin to appreciate scholarship as activism and recognise activism as scholarship. In this respect, scholars are increasingly working in dialogue with alternative forms of scholarship and practice that cross and deconstruct boundaries of discipline, nation-state, and mobilities. Msezane's installation is one of many in contemporary South Africa and many global contexts, where there is an increasing call for embodied, affective, and careful research. There are many further novel and creative intersectional gender justice moments, actions, and movements, often drawing on activist and activist-artistic modes, or artivism (Nossel, 2016). In global contexts, an inspiring example is walking activist-artistic projects, for example, the *Walk* in the city of Delhi to protest gender-based violence was generated in India by a performance company following the sexual violation and murder of Jyoti Pandey (Arora, 2020). Such engagements begin to build and bolster a proliferating archive of activist-knowledge and knowledge-activism that offers powerful interpretative resources for those engaged in justice knowledge.

Qualitative research that embraces and engages thinking with/through/in overlapping folds of art, activism, and research also transgresses disembodied and dis-affected/affective scholarship, in which the mind and rationality reign. Such embodied, affective scholarly projects and situated knowledges resist representational research, foregrounding a relational ontology, recognition, and an ethics of care, both materially and symbolically. Collaborating with activist-performance and other activisms in our pedagogies and research is a powerful way of transgressing the rigid normativities and violences of the academy and the colonial and patriarchal repetitions of our scholarship. Alaimo (2016, p. 112) points out:

> Activists, as well as everyday practitioners of environmental, environmental health, environmental justice, and climate change movements, work to reveal and reshape the flows of material agencies across regions, environments, animal bodies, and human bodies – even as global capitalism and the medical-industrial complex reassert a more convenient ideology of solidly bounded, individual consumers and benign, discrete products.

One emergent productive way of deploying embodied and affective interpretations, through art and activism, while also avoiding extractive practices, is through what has increasingly been termed 'research-creation' (Manning, 2016; Truman & Springgay, 2016; Truman, 2022): a participatory transdisciplinary collaboration between those with different knowledges working together in creative ways for knowledge that is of value in both the process and product.

A wide range of research-creation projects are currently flourishing (Truman, 2022), albeit still in largely global northern privileged spaces, but also in virtual spaces (e.g., *Feral Atlas,* created by Tsing et al., 2021, https://feralatlas.org/) that are creating hubs of possibility for knowledge-activism and activist-knowledge. Such forms of justice (post)qualitative research both disrupt the binaries of mind-body, knowledge-activism, and so on and also open up response-able engagements since the research-creation is read also as a response to injustice that may make a difference in policy and practice. In the speeded-up neoliberal current of the post-academic university (Donskis, 2019) and the global state of the emerging ecological crisis, it becomes more imperative to pursue with urgency these creative, playful, disruptive, and political spaces for critical interpretation.

References

Alaimo, S. (2016). *Exposed: Environmental Politics and Pleasures in Posthuman Times*. University of Minnesota Press.

Arora, S. (2020). Walk in India and South Africa: Notes towards a decolonial and transnational feminist politics. *South African Theatre Journal*, 33(1), 14–33.

Buikema, R. (2021). *Revolts in Cultural Critique*. Rowman & Littlefield.

Donskis, L. (2019). Toward an educational dystopia? Liquid evil, TINA, and post-academic university. In L. Donskis, I. Sabelis, F. Kamsteeg, & H. Wels (Eds.), *Academia in Crisis: The Rise and Fall of Neoliberal Education in Europe* (pp. 11–35). Brill.

Hemmings, C. (2012). Affective solidarity: Feminist reflexivity and political transformation. *Feminist Theory*, 13(2), 147–161.

Hubbard, P. (2009). The Zimbabwe birds: Interpretation and symbolism. *Honeyguide*, 55(2), 109–116.

Lalu, P. (2023). *Undoing Apartheid*. Polity Press.

Lenz Taguchi, H. (2010). Images of thinking in feminist materialism: Ontological divergences and the production of research subjectivities. *International Journal of Qualitative Studies in Education*, 23(5), 525–542.

Manning, E. (2016). *The Minor Gesture*. Duke University Press.

Murris, K. (2016). #Rhodes must fall: A posthumanist orientation to decolonising higher education institutions. *South African Journal of Higher Education*, 30(3), 274–294.

Nagar, R. (2013). Storytelling and co-authorship in feminist alliance work: Reflections from a journey. *Gender, Place & Culture*, 20(1), 1–18. https://doi.org/10.1080/0966369X.2012.731383.

Nagar, R. (2014). *Muddying the Waters: Coauthoring Feminisms Across Scholarship and Activism*. University of Illinois Press.

Nagar, R., in journeys with Sangtin Kisan Mazdoor Sangathan, & Parakh Theatre. (2019). *Hungry Translations: Relearning the World Through Radical Vulnerability*. University of Illinois Press.

Nossel, S. (2016). Introduction: On "artivism," or art's utility in activism. *Social Research: An International Quarterly*, 83(1), 103–105. https://doi.org/10.1353/sor.2016.0023.

Pather, J., & Boulle, C. (Eds.). (2019). *Acts of Transgression: Contemporary Live Art in South Africa*. Wits University Press.

Romano, N. (2019). Just(ice) do it! Re-membering the past through co-affective aesthetic encounters with art/history. *Alternation*, 26(2), 62–88.

Sangtin Writers Collective & Nagar, R. (2006). *Playing with Fire: Feminist Thought and Activism Through Seven Lives in India*. University of Minnesota Press.

Shefer, T. (2018). Embodied pedagogies: Performative activism and transgressive pedagogies in the sexual and gender justice project in higher education in contemporary South Africa. In V. Bozalek, R. Braidotti, T. Shefer, & M. Zembylas (Eds.), *Socially Just Pedagogies in Higher*

Education: Critical Posthumanist and New Feminist Materialist Perspectives (pp. 171–188). Bloomsbury.
Spivak, G. (1988). Can the subaltern speak? In C. Nelson & L. Grossberg (Eds.), *Marxism and the Interpretation of Culture* (pp. 271–313). University of Illinois Press.
Truman, S. E. (2022). *Feminist Speculations and the Practice of Research-Creation: Writing Pedagogies and Intertextual Affects*. Routledge.
Truman, S. E., & Springgay, S. (2016). Propositions for walking research. In P. Burnard, E. Mackinlay, & K. Powell (Eds.), *The Routledge International Handbook for Intercultural Arts Research* (pp. 259–267). Routledge.

2 Architecture

Randall Bird

Architecture Through Interpretation

Like cuisines around the globe, architecture consists of interpretative ideas and practices that are in constant motion and ever-changing. In their book, *Interpretation in Architecture: Design as a Way of Thinking*, Adrian Snodgrass and Richard Coyne argue that interpretation is ubiquitous in architecture, both as critique (understanding) and in the process of architectural creation (production) (Snodgrass & Coyne, 2005). On the one hand, conversing, thinking, or writing about architecture, that is, the forms of interpretation, requires a transformation of the visual into the verbal. On the other hand, the act of designing architecture, whether in an academic studio or in a busy office, involves a creative act of interpretation by giving shape to new forms and assembling materials in specific ways. In the history of architecture, there are countless approaches to 'interpretation' by academics and practicing architects, so I will only scratch the surface while providing a few examples.

Interpretation in architecture, at its core, is relational, by and through which its parts are related to the whole and vice versa. In his brilliantly concise essay 'Structure, Construction and Tectonics', the late architectural historian Eduard Sekler elaborates on this point by examining the relationship among these key architectural aspects (Sekler, No Date). He argues that 'construction' implies putting something together or an assemblage of parts, whereas 'structure' connotes a deliberate arrangement of various material elements or pieces to deal with the forces at play in any form of architecture. Insightfully, Sekler points out that 'tectonics' is that relational quality that provides structure and construction with a visual expression that may arouse feelings of, for example, empathy, inspiration, reverence, or pleasure. Tectonics is the static expression, he argues, of what brings life to architecture and therefore is at the center of how we assess or interpret it. Following Sekler's lead, it is important for the qualitative researcher to be attentive to the expressive and emotive aspects of architecture.

DOI: 10.4324/9781003585046-3

Sekler's key notions have deep roots in the ideas of the Roman soldier and architect Vitruvius (d. ca. 15 BCE), who is among the earliest Western architectural theorists and interpreters. Vitruvius puts forth criteria for the means by which a building's quality can be judged or interpreted. Drawing from ancient Greek classical architecture, he argues that a building must not only be built solidly but also must fulfill its purpose, or its function, including its beauty (Vitruvius, circa 30 BCE). For Vitruvius, embedded in the notions of function and beauty is symmetry or the balanced proportional relationship of individual architectural elements (e.g., columns) to the whole. He emphasizes proportion in architecture because it is the relationship between parts and the whole that provides a building with a harmonious quality, which is the most important consideration when interpreting the merit of a building. Reflecting on Vitruvius (and Sekler), qualitative research in the realm of architecture is enhanced and furthered by close analysis of how the whole is greater than the sum of its parts.

Beyond Western architectural examples, casting a wide net, on the part of the qualitative researcher, is an integral process for interpreting layers of historical and contemporary architectural meaning and building practices. In the central highlands of Madagascar, qualitative research by necessity involves interpreting the siting, orientation, and design of tomb and domestic architecture and involves a critical examination of how architecture, landscape, and the sky are seen in relation to one another (Bird, 2012). Important features of Malagasy cosmology are made concrete by visual metaphors mapped out on the floor surfaces of houses and tombs. For example, the 28-day lunar cycle is referenced at specific points where the horizontal floor meets the corners and vertical walls in tombs and houses. The northeast corner, which is associated with the first day of the lunar cycle, is the most auspicious and is associated with sacred reverence for ancestors. The result of symbolic and imaginary points in a cosmological cycle is realized in physical form. In present-day Malagasy homes, the northeast corner is adorned with portraits of family ancestors and various offerings, including porcelain, silver, and flowers. The entrance to a house is traditionally located at the southwest corner. Upon entering a house, there is an immediate visual connection to the opposite sacred northeast corner, thereby creating a metaphorical link between the living and the dead.

Subject to interpretation, other architectural structural features of the house carry layers of significance beyond their obvious function. In the center of the floor plan is a central pillar that extends vertically from the floor to the main beam of the roof above. At its most basic level of interpretation, the trabeated system of construction consists of a vertical pillar supporting a horizontal beam and acts to counteract the force of gravity (Bird, 2012). Extending interpretation to include symbolic and cosmological aspects, the central pillar represents the place – the still point of the house – where 'destiny' converges. 'Destiny' is a power that influences every aspect of life events (birth, marriage, death, planting of crops, and construction of a house or a tomb). The central pillar supports the roof and metaphorically brings life, safety, and good fortune to the family as the core and heart of a house. The pillar doubles as a symbolic and physical connection to the portions of 'destiny' that remain invisible – those marking periods of darkness beneath the ground on which the house is built.

The ground on which houses and tombs are constructed and, by extension, the wider landscape are at the very core of Malagasy identity (Bird, 2012). The attachment to a place, both at micro- and macroscales, is central to the construction and maintenance of individual and group identity as generations of Malagasy draw on the landscape as the main source of connections to their ancestors. Seen from the vantage point of the living, the landscape is the ultimate source of belonging because people connect to one another by and through engaging with their ancestors (Bird, 2012). Most Malagasy refer to landscape as 'a land of the ancestors', which is a distillation and summation of the extreme importance of the relationship between the living and the dead, land, sky, and architecture (e.g., tombs and houses). Ancestral land and the architecture constructed on the land are tantamount to one's own status in society and historical and current affiliation with other Malagasy, thereby creating a critical sense of belonging.

In his recently published *On Bramante*, Pier Paolo Tamburelli argues that architecture, an intellectual project, is embedded with 'an unsuspected power of criticism, almost as if the loss of an explicit mission assigned to it by society had made it strangely freer than in the past' (Tamburelli, 2022, p. XI). Accordingly, qualitative research is driven by interpretation as critique, as tectonics, as empathy, as assessment, and as engendering layers of historical and cultural understanding beyond what is apparent in physical form and function. This approach is crucial to how architecture is thought about, produced, and experienced across time in an ever-changing environment.

References

Bird, R. (2012). Earth, sky, ancestors: Building the merina world in nineteenth-century Madagascar. In *African Cosmos: Stellar Arts*. Washington, DC: National Museum of African Art, Smithsonian Institution.

Sekler, E. F. (n.d.). Structure, construction & tectonics. In Gyorgy Kepes (Ed.), *Harvard Graduate School of Design*. According to the author, the article is a slightly altered version of one written in 1963

and published in *Structure in Science and Art* (pp. 3–11). New York: George Braziller, 1965.
Snodgrass, A., & Coyne, R. (2005). *Interpretation in Architecture: Design as a Way of Thinking*. Routledge.
Tamburelli, P. P. (2022). *On Bramante*. The MIT Press.
Vitruvius, (circa. 30 BCE). (1960). *De architectura* (Trans. Morris Hickey Morgan). Rome as *The Ten Books of Architecture*. Cambridge, MA, 1914; reprint New York.

3 Collective Biography

Mnemo ZIN, Susanne Gannon, and Outi Ylitapio-Mäntylä

Collective Biography in Education Research

Collective biography is an approach to memory work that foregrounds a collective process of remembering and often focuses on childhood memories. This is because the earliest traces of subject formation – of the discourses, practices, rationalities, and socio-material-relational contexts within which we come to take ourselves up or recognize ourselves in particular ways – can be traced to childhood. Instead of focusing on the psychological or individual subject, collective biography examines processes of subjectification within socio-material worlds by bringing together a group of people to investigate an agreed-upon topic through their memories. In educational research, the topics of concern may be obliquely or directly associated with educational practices or everyday childhood experiences, which are particularly generative for adult educators.

Collective biography developed from the pioneering work of critical feminist sociologist and philosopher Frigga Haug and the collective of women in Germany who came together in the 1980s to investigate the sexualization of women's bodies (1983/1987). Haug and the collective called their process "collective memory work" and were committed to "reconstructing scientific work along feminist lines" against the limits of conventional sociology and psychology (Haug et al., 1987, p. 23). Their goal was to address the gap between Marxist theory and their own experiences as women through memories in order to examine the ways in which women "enter into pre-given structures, within which they reproduce both themselves, and the categories of society" (Haug et al., 1987, p. 40). Although originally collective memory work was deeply rooted in Marxist feminism (Haug, 2003), it has been later adapted for educational research overtly drawing on feminist post-structural theories and new materialisms (Davies & Gannon, 2006; Gannon & Davies, 2021; Gonick & Gannon, 2014) and decolonial thought (Millei et al., 2018; Silova et al., 2017; Zin & Gannon, 2022).

DOI: 10.4324/9781003585046-4

The approach to collective biography developed by Haug and the original collective was framed as "a transgression of boundaries" (Haug et al., 1987, p. 36), stemming from their insistence on working as a collective of women rather than as individuals and bringing what had been private matters (the body, the family, girlhood, and domestic lives) out into the open for critical interrogation. While reclaiming their power and agency, Haug's collective saw the transgressive potential of collective memory work in effectively challenging – and ultimately altering – the existing status quo along with the established knowledge traditions and subject positions.

Transgression is a valuable lens for thinking through how collective biography challenges accepted methodological practices, orders, and norms. In mainstream educational research, the sorts of transgressions enabled by collective memory work involve crossing the established divides between past and present, subject and object, self and other, individual and collective, mind and body, and emotion and reason. In particular, "working through" memories is an active reconciliation of past and present together with the tensions and emotions built and bottled up within those. Memory stories bring our changing sense of who we are and who we were into view, often disrupting habituated ways of telling and thinking about our lives and *selves*. They may elicit various affective responses, including complex combinations of relief, regret, pleasure, or ambivalence. It is these points of discomfort, contradiction, and ambiguity that are often most productive analytically when working with memories because they disrupt both scientific and political master narratives.

Methodologically, collective biography projects do not follow formulaic or replicative processes; rather, they rely on the elasticity and adaptability of these modes of research depending on the goals, characteristics, and configurations of participants in each collective. While some memory-work approaches have been more explicit in laying out "rules" and outlining specific research phases, steps, and templates (e.g., Crawford et al., 1992; Onyx & Small, 2001; Hamm, 2021), others have emphasized more open-ended, emergent strategies and workshop designs adapted to each particular inquiry. For example, Davies and Gannon (2006) insist that collective biography entails emergent and situated practices in terms of the ways in which research groups may choose topics, invoke memories, or incorporate other modalities (e.g., images, objects, art-making, theatre, and bodywork). Their touchstone is Haug's insistence that "there might well be no single, 'true' method that is alone appropriate to this kind of work . . . the very heterogeneity of everyday life demands similarly heterogeneous methods" (Haug et al., 1987, pp. 70–71).

There are also variations in how each group goes about writing and analysing memories (see Haug et al., 1987; Davies & Gannon, 2006;

Gonick & Gannon, 2014; Millei et al., 2019). While Crawford et al. (1992) and Onyx and Small (2001) suggest that participants bring an already written memory to a workshop, Davies and Gannon (2006) and Mnemo Zin and Gannon (2022) see the process of telling stories to the group out loud first before they are written down as important to allow the dynamic and resonant affective embodied space of the workshop itself to be more fluid and generative. Reflexive listening offers the potential for listeners to enter imaginatively into the memory being shared, to feel its resonances, blind spots, and gaps.

Certain practices have been consistent across all versions of these methods, including sharing a memory of a discrete episode or moment from one's experience, writing it in the third person, and including as much detail as possible (taste, smell, and sound) to help listeners imagine the material setting for the memory (time, light, objects, other people, etc.). Delving into these concrete details helps the author retrieve an embodied sense of how it was to be there, in that moment. Through the phases of a workshop, the participants take turns sharing their memories while asking questions, clarifying details, identifying points of similarity and difference, and noting clichés, stereotypes, and generalizations. While the memory storytellers should make the scene as vivid as possible, they should not explain or interpret the memory at this stage. Collective interpretation, analysis, and writing follow next.

In most collective biography research, analysis continues after a workshop has finished through a long process of collaborative authorship, as theoretical resources continue to build as needed and as collective biography researchers continue to think and write together. Although different scholars recommend various (sometimes specific) approaches to further analysis, we suggest that researchers develop bespoke ways of interpretation that best fit *their* particular inquiry. If members of the collective share an interest in a particular concept, they may gather key readings to guide analysis, drawing on memory stories as data to work theory into the everyday (Davies & Gannon, 2006). Members of the collective may decide to first write down their own interpretations of their memory stories and then share their reflections with others. As the interpretation of the memory story passes from author to author, the analysis becomes more layered and complex. Through all stages of analysing, drafting, exchanging, and revising, each paper may go through 12–14 versions of rewriting. During this process, it is possible that the collective of researchers may lose sight of where or who each story came from or who contributed which line of analysis and interpretation. The writing process includes false starts, leaps forward, tangents, and continuous wrestling with ideas collectively all the way through.

Collective biography has inspired educational research in many areas. Researchers have examined feminist educators' relations with the natural world (Kaufman et al., 2006), teenage girls' relationship with their bodies and sexuality (Oinas, 2001; Gonick & Gannon, 2014), heteronormativity, gender, and sexuality in school practices (Purohit & Walsh, 2003), gendered craft education (Kokko, 2009), adult learning (Wichmand & Koelbaek, 2021), early childhood and care (Ylitapio-Mäntylä, 2013), refugee education (Fiedler, 2021), postsocialist childhood and schooling (Millei et al., 2018, 2019; Mnemo Zin & Gannon 2022), for teaching (Clift & Clift, 2017; Ylitapio-Mäntylä & Mäkiranta, 2010), or academic labour (Davies et al., 2005; Gannon et al., 2019). In each of these different contexts, collective biography is reinvented in new ways, as each new group of researchers, writers, and artists works together to explore issues of common concern in particular times and spaces.

References

Clift, B.C., & Clif, R. T. (2017). Toward a "Pedagogy of Reinvention": Memory Work, Collective Biography, Self-Study, and Family. *Qualitative Inquiry*, 23(8) 605–617.

Crawford, June, Kippax, Susan, Onyx, Jenny, Gault, Una, & Pam, Benton. (1992). *Emotion and Gender: Constructing Meaning From Memory.* Sage.

Davies, B., Browne, J., Gannon, S., Honan, E., & Somerville, M. (2005). Embodied Women at Work in Neoliberal Times and Places. *Gender, Work and Organization*, 12(4) 343–362.

Davies, B., & Gannon, S. (Eds.). (2006). *Doing Collective Biography.* Open University Press.

Fiedler, B. (2021). Collective Memory Work with Refugee Women Who Don't Have a Common Language. In R. Hamm (Ed.), *Reader: Collective Memory-Work* (pp. 482–489). Beltra Books. https://collectivememorywork.net/wp/wp-content/uploads/2024/12/Reader-CMW-ebook-complete-revised.pdf

Gannon, S., & Davies, B. (2021). Doing Collective Biography Through Theoretical, Material, Affective, Embodied Assemblages. In R. Hamm (Ed.), *Reader: Collective Memory-Work E-Book* (pp. 540–569). Beltra Books.

Gannon, S., Taylor, C., Donoghue, H., Harris-Evans, J., Healey, J., Hannam-Swain, S., Moore, P., & Adams, G. (2019). 'Working on a Rocky Shore': Micro-Moments of Positive Affect in Academic Work. *Emotion, Space and Society*, 31 48–55.

Gonick, M., & Gannon, S. (2014). *Becoming Girl: Collective Biography and the Production of Girlhood*. The Women's Press.

Hamm, R. (2021). *The Potential of Collective Memory-Work as a Method of Learning: Application and Adaptations*. Beltra Books.

Haug, F., Andresen, S., Bünz-Elfferding, A., Hauser, C., Lang, U., Laudan, M., Lüdermann, M., & Meir, U. (1983). *Sexualisierung: Frauenformen 2. Argument.* [this is the original German version of the book below translated into English]

Haug, F. (1987). *Female Sexualization: A Collective Work of Memory*. Verso.

Haug, F. (2003). Marx Within Feminism. Talk Given at the Complutense University of Madrid, Spain. [Online]. Available: http://friggahaug.inkrit.de/documents/marxinfem.pdf.

Kaufman, J. S., Ewing, M. S., Hyle, A. E., Montgomery, D., & Self, P. A. (2006). Women and Nature: Using Memory-Work to Rethink our Relationship to the Natural World. *Environmental Education Research*, 12(3–4) 309–326. https://doi.org/10.1080/13504620600942774

Kokko, S. (2009). Learning Practices of Femininity Through Gendered Craft Education in Finland. *Gender and Education*, 21(6) 721–734. https://doi.org/10.1080/09540250903117090.

Millei, Z., Piattoeva, N., Silova, I., & Aydarova, E. (2018). Hairbows and Uniforms: Entangled Politics in Children's Everyday Lives. In I. Silova, N. Piattoeva, & Z. Millei (Eds.), *Childhood and Schooling in (Post)socialist Societies: Memories of Everyday Life* (pp. 145–162). Palgrave Macmillan.

Millei, Z., Silova, I., & Gannon, S. (2019). Thinking Through and Coming to Know With Memories of Childhood in (Post)Socialist Spaces. *Children's Geographies*, 20(3) 324–337.

Millei, Z., Silova, I., & Piattoeva, N. (2018). Towards Decolonizing Childhood and Knowledge Production. In I. Silova, N. Piattoeva, & Z. Millei (Eds.), *Childhood and Schooling in (Post)socialist Societies: Memories of Everyday Life* (pp. 231–256). Palgrave Macmillan.

Mnemo Zin & Gannon, S. (2022). Scenes from a collective biography of Cold War childhoods: A decolonial ethnodrama. *Cultural Studies – Critical Methodologies*, 22(3) 235–244.

Oinas, Elina 2001. *Making Sense of the Teenage Body: Sociological Perspectives on Girls, Changing Bodies and Knowledge.* Abo Akademi University Press.

Onyx, J., & Small, J. (2001). Memory-Work: The Method. *Qualitative Inquiry*, 7(6) 773–786.

Purohit, K. D., & Walsh, C. (2003). Interrupting Discourses Around Gender Through Collective Memory Work and Collaborative Curriculum Research in Middle School. *Sex Education: Sexuality, Society and Learning*, 3(2) 171–183.

Silova, I., Millei, Z., & Piattoeva, N. (2017). Interrupting the Coloniality of Knowledge Production in Comparative Education: Postsocialist and Postcolonial Dialogues After the Cold War. *Comparative Education Review*, 61(S1) S74–S102.

Wichmand, M., & Kolbaek, D. (2021). Collective Memory Work: A Method for Turning Adult Learners' Work-Life Experiences into a Rich Collective Knowledge Resource in Higher Education. *Adult Learning*, 10451595211004767.

Ylitapio-Mäntylä, O. (2013). Reflecting Caring and Power in Early Childhood Education: Recalling Memories of Educational Practices. *Scandinavian Journal of Educational Research*. Available: http://www.tandfonline.com/doi/abs/10.1080/00313831.2011.637230.

Ylitapio-Mäntylä, O., & Mäkiranta, M. (2010). From Past to Present: Photo-Related Memory-Work and New Images in Educational Practices. In M. Bielecka (Ed.), *Tradition and New Practices in Art Education* (pp. 119–126). Jan Kochanowski University of Humanities and Sciences, Ministry of Science and Higher Education.

4 Community of Philosophical Enquiry

Joanna Haynes, Karin Murris, and Rose-Anne Reynolds

The Community of Philosophical Enquiry Research Methodology

The community of philosophical enquiry (CoPE) research methodology resembles, at a superficial level, that of research approaches such as 'focus groups' or other small group interventions. Despite these surface similarities, the energy and dynamism of a CoPE are furnished through the asking of open-ended philosophical questions. This changes what is meant by interpretation: *who* interprets, *how* interpretation happens and *what* the focus is of the interpretation. By putting open-ended questioning at the heart of research enquiry with all its uncertainty and unpredictability, even what counts as data is at stake. Research participants think together and shape and reshape the core concepts implied in the investigation.

The CoPE is now adopted in a variety of academic and other settings like community work, informal learning, social policy, lifelong education, early childhood education, and childhood and youth studies. Its relevance for research purposes arises directly out of its transformative power as a pedagogy (see, e.g., Murris, 1997; Haynes, 2002/8; Tiffany, 2009; Murris, 2016; Murris & Haynes, 2018; Reynolds, 2021; Giorza, 2021; Crowther, 2021). In developing the community of enquiry pedagogy in the Philosophy for Children programme, Ann Margaret Sharp and Matthew Lipman revised Charles Peirce's notion of a community of experts into a pedagogical device, prompted by Dewey's democratization of communal inquiry as a method of problem-solving among non-experts (Gregory, 2022). Thus, CoPE is a deep theory of collective knowledge generation and its adoption has a profound impact on the interpretation of knowledge and research data.

In the CoPE, it is not only the theory of enquiry that is democratizing, but learning *about* the method involves a participatory and democratic ethic of shared interpretation – a *doing* of philosophical enquiry *with others*. It is a challenging and provocative approach whose openness

DOI: 10.4324/9781003585046-5

creates spaces for the exposure of new possibilities, notwithstanding the necessity for alertness and attention to relations of authority and control. Involvement in communities of philosophical enquiry raises profound questions about subjectivities, agency, knowledge and ethics. Researching through the CoPE is an active process of intervention in which the researcher is deeply implicated. Therefore, interpretation is not something that happens after a data harvest or later in a research project but is always already happening. For researchers committed to working in egalitarian ways that break down barriers and promote collaboration and dialogue as part of the research process, the CoPE offers an approach that can help to realize such values.

Unlearning How to Question

In education generally, most of the questions asked are of a *factual* or *empirical* nature, answerable by 'look-and-see' or using one's senses, including scientific experiments. They are *closed questions* in the sense that they are information-seeking or even rhetorical. These questioning practices are also evident in many research practices, in which the research often starts by identifying a problem or a gap and articulating related research questions. These questions then inform the research design and involve choosing interventions/methods to help answer them and, accordingly, to interpret the data, in line with the 'original' questions and aims.

By contrast, questioning and interpretation in the CoPE involve *unlearning* how we have learnt to question in education or more traditional forms of research and assume a very different kind of listening and interpretation by the researcher. Training in the approach is essential.[1] Who gets to ask the questions is not only a political question (Benjamin & Echeverria, 1992); different and more wide-ranging research data are produced when the CoPE is adopted for research. Open philosophical questions resist interpretive closure through single answers – they are *divergent* and open up further questioning. Kennedy and Kohan (2008, p. 9) argue that 'a question is interesting not so much because of what it is or it might be, but because of the movement that it can generate in the questioner and the questioned'.

Thinking, listening and dialoguing together in CoPE complexify the interpretation of the issue at hand by opening up the use of a project's *core* concepts, such as 'community', 'research', 'technology' or 'play'. These are examples of common, everyday concepts and, as such, are central in the language we use to think with (together) in research practices. The abstract nature of such philosophical concepts makes them contestable, as their meaning depends on the context in which they are used, and they are always open to further questioning. Golding (2015, p. 208) suggests that philosophical questions cannot be resolved by empirical methods because the issue is a conception that needs further clarification, rather than a 'gap' in our empirical knowledge. He argues that we solve the problem by enriching our conceptions, or by refining them, to make them congruent with other conceptions or with our experience. By pursuing further enquiry and asking probing open philosophical questions, concepts are reshaped by focusing on the particularity of the enquiry. This connection between the theoretical and the experiential, and the concrete and the abstract aids a deeper understanding and interpretation of the issues to be investigated.

It is as a result of the significant shift in *who* usually asks the questions, in the *philosophical* rather than empirical nature of the questions and in the *dialogical* character of the process of the CoPE that the space is made for a different kind of listening by *all* participants, including the researcher. It has the potential to create more egalitarian relationships between research participants. The normalization of disagreement in the CoPE expresses the intellectual and affective value of doubt and uncertainty, and it articulates a provisional relationship to knowledge and truth that is unusual in most interpretive practices.

Questioning, Listening and Interpreting in a Community of Philosophical Enquiry

- The gathering of the group is circular, with everyone facing each other and at the same level, emphasizing equality of interpretive status;
- Processes for enquiries are discussed and reviewed, including procedures for turn-taking and challenging the process itself;
- An ambiguous provocation is shared to invite questioning from all participants;
- Questions are generated from within the group – there is more questioning than answering;
- Enquiries are concept-driven; key concepts are problematized, shaped and reinterpreted through the course of the enquiry with the help of examples from lived and imagined experiences;
- The researcher acts as a facilitator who works to create and sustain good conditions for enquiry and keeps putting concepts back into question for further interpretation;
- The researcher facilitates the making of connections, building on ideas that emerge and promoting deep dialogue;
- Listening is open-ended; the focus is on attentive noticing and listening *to* the community; the process disrupts hierarchies of authority – interpretive authority is distributed;
- Participants changing their minds and unexpected answers are commonplace – and desirable; contestation disagreement is a normalized part of interpretation;

- Participants engage together in fresh exploration that comes about through the sharing of thinking – positions change – new questions arise;
- Enquiries spiral, rather than being linear diverge, are open-ended and philosophical.

Researchers adopt these ethical, ontological and epistemological principles to shape the approach to interpretation and interpretive practices.[2]

Notes

1 For training options consult in the UK (https://www.sapere.org.uk/our-courses/) or the global network (https://www.icpic.org/); see also https://karinmurris.com/p4c/.

2 For a more expanded version of this article, see Haynes and Murris (2025).

References

Benjamin, M., & Echeverria, E. (1992). Knowledge in the classroom. In A. M. Sharp & R. F. Reed (Eds.), *Studies in Philosophy for Children Harry Stottlemeier's Discovery* (pp. 64–78). Temple University Press.

Crowther, J. (2021). *'Undergoing' as Posthuman Literacy Research in an In/Formal Settlement Primary School in South Africa*. Unpublished Doctoral Thesis, University of Cape Town.

Giorza, T. (2021). *Learning With Damaged Colonial Places: Posthumanist Pedagogies from a Joburg Preschool*. Springer.

Golding, C. (2015). The community of inquiry: Blending philosophical and empirical research. *Studies in Philosophy and Education*, 34, 205–216.

Gregory, M. (2022). Charles Peirce and the community of inquiry. *Analytic Teaching and Philosophical Praxis*, 42(1), 1–16.

Haynes, J. (2002, 2008). *Children as Philosophers: Learning Through Enquiry and Dialogue in the Primary Classroom*. 2nd Edition. Routledge.

Haynes, J., & Murris, K. (2025). Interpretation and intra-pretation in communities of philosophical enquiry: A post/qualitative research methodology. In M. Koro & K. Murris (Eds.), *Handbook of Reconfiguring Interpretation in Postqualitative Research*. Routledge.

Kennedy, D., & Kohan, W. O. (2008). Aión, kairós and chrónos: Fragments of an endless conversation on childhood, philosophy and education. *Childhood & Philosophy*, 4(8), 5–22.

Murris, K. (1997). *Metaphors of the Child's Mind: Teaching Philosophy to Young Children*. Unpublished Doctoral Thesis, Hull University, United Kingdom.

Murris, K. (2016). *The Posthuman Child: Educational Transformation Through Philosophy With Picturebooks*. Routledge.

Murris, K., & Haynes, J. (Eds.). (2018). *Literacies, Literature and Learning: Reading Classrooms Differently*. Routledge.
Reynolds, R. (2021). *A Posthuman Reconfiguring of Philosophy With Children in a Government Primary School in South Africa*. Unpublished Doctoral Thesis, University of Cape Town.
Tiffany, G. (2009). *Community Philosophy: A Project Report*. Joseph Rowntree Foundation. https://www.jrf.org.uk/report/community-philosophy-project-report.

5 Criticality

Vivienne Bozalek and Karen Malone

Critical Interpretation in Postqualitative Research

In conventional qualitative research, interpretation as criticality is seen as the quality or state of being critical and is very much bound up with reflexivity or critical reflection. "Criticality in interpretation" as a term is used traditionally to question how a researcher has engaged with the "evidence" (often viewed as facts or truths) and how this engagement has informed and been demonstrated through their understanding. What influenced your findings, and how did you come to your conclusions? Some time ago, Bruno Latour (2004) proposed that critique had run "out of steam" and in his article on this topic posited the question:

> Can we devise another powerful descriptive tool that deals this time with matters of concern and whose import then will no longer be to debunk but to protect and to care, as Donna Haraway would put it? Is it really possible to transform the critical urge in the ethos of someone who adds reality to matters of fact and not subtract reality?
>
> (p. 232)

That is, by "acknowledging the agency of the world in knowledge makes room for some unsettling possibilities" (Haraway, 1988, p. 593), Latour proposed the question, "can the tool of criticism be brought to critique itself?"

For Haraway (2016), asking questions means that there is no presumption about what pre-exists; in research, what is required is cultivating a sensibility of curiosity. She refers to Vinciane Despret's practice of crafting polite questions, which holds open the possibility that one might be surprised by something interesting which might happen, which changes everyone involved, in a "subject- and object-making dance" (Haraway, 2016, p. 127). Haraway (2016) notes how such a practice of posing and responding to interesting questions is similar to training the mind and imagination to go visiting, as proposed by Hannah Arendt and Virginia

DOI: 10.4324/9781003585046-6

Woolf. Criticality for Haraway means being for some worlds and not for others and moving beyond this to help compose those worlds.

Interpretation doesn't happen where we assume it does – past, present and future all affect interpretation, which is constituted by all that is gone but which is still here and will continue to be here (Manning, 2020). Interpretation is immanent to the event and is carried by it, rather than being based on the assumption that there is some sort of collective understanding of critique, especially of artistic practice. Such critique and the interpretation that comes from it are never neutral but based on neurotypical and white habitual presuppositions and other political ways of viewing the world, which overcode what is seen, rather than attuning to what is under scrutiny (Manning, 2020).

Viewing work such as a text or a visual art performance or piece from the outside, or relating it to the critic's own work or a lineage of previous work, rather than an ethos of encounter with what moves the work from within, can be seen as a form of debased interpretation. This is where the immanent field of value is neutralised and made redundant. Here, distance as critique is not critiqued but taken for granted. For interpretation to be of value, otherness needs to be included, beyond the habits of perception.

Questions from the view of critique as distance tend to be deadening in that the answers are already known – not like Despret's polite and curious forms of questioning. Schizzing, according to Manning, allows us to move beyond the habitual and the idea that we already know. Paying attention to what would otherwise have remained muted and invisible potentiates a speculative sharing of what it might generate. This might happen, for example, through asking a new set of questions in order to discover the potential in the constellation of the work. Like Haraway and Barad, who eschew critique as judgement, Manning's schizoanalysis does not compare or judge against pre-existing societal norms but invents intercessing techniques, which occur at the pre-individual level and shift the relational field. This generates ways of retooling critique into immanent critique, where interpretation is a process from within, "inquiring how the work does its work" (Manning, 2020, p. 112). In this case, interpretation can be experimental, creative and affirmative, potentiating forms of inquiry that make a difference.

Extending these ideas, a diffractive methodology provides different optics to the tools of critical reflection or reflexivity (Haraway, 2016; Barad, 2007). Critical reflection or reflexivity has been used in research for conducting a comparison between two or more pre-existing theories or positions. From a relational ontology, however, individual entities do not pre-exist in their encounters but come into being through relationships. This implies that two theories or positions cannot be compared because they always already exist inside each other and do not pre-exist their encounters with each other. Because critique (of something) is closely linked to Cartesian and Newtonian representationalist habits of thinking, which presume the existence of individual entities occupying places in the world, a relational way of thinking seems unimaginable. A relational ontology relies on the complexity of encounters where entanglements rather than objects are the focus of thinking.

Interpretation as intra-active rather than interactive shifts the focus of attention from the engagement of individual acting agents and looks to the in-between agency of the encounter.

Barad's (2007) theorising through agential realism sees politics, ethics, epistemology and ontology as being inextricably linked. Conventional notions of ontology within this framing are undone – entities or things are not seen as given or as fixed states but are dynamic and indeterminate. Critical interpretation in this dynamism means there is an ongoing radical openness with an incalculable infinity of possibilities for reconfiguring the conditions in which the world is materialised in certain ways and not in others. Thus, interpretation as a criticality in agential realism requires much more – it involves ways of thinking thought otherwise (Barad & Gandorfer, 2021). Interpretation as criticality within agential realism disrupts an analysis from a distance, as there are no determinate boundaries and no bounded individual subjectivities that pre-exist and are separable from what is being investigated. Critical interpretation or sense-making here would mean getting underneath thought. One way of doing this would be by tracing the entanglements of material and historical conditions in which particular ways of thought have come to matter and then opening these up to reworking. Inhabiting such interpretative criticality would require asking questions through employing agential cuts.

Agential cuts in this sense are what Barad refers to as "cutting-together-apart" or "differentiating-entangling". Intra-acting as emergent critical interpretation discerns how the material-discursive conditions under which thought become sedimented in particular ways and not others, for as Barad notes, "intra-actions enact agential cuts, which do not produce absolute separations, but rather cut together-apart (one move)" and consequently, "there is no moving beyond, no leaving the 'old' behind. There is no absolute boundary between here-now and there-then. There is nothing that is new; there is nothing that is not new" (Barad, 2014, p. 168). Interpretation as criticality involves breaking out of binarised ways of thinking by undoing and interrupting cuts such as here/there, now/then, subject/object, material/discursive, word/world, matter/meaning and understanding these in their inseparability. An important part of interpretative criticality in agential realism is an understanding of justice-to-come which is an ongoing, open ethical practice in that it will never finally arrive. Nothing is presumed as being familiar, and no end-point is anticipated. Criticality is a practice rather than a judgement.

Sedimentations of materialisation are not seen as given but as historicised in specific iterative intra-actions. By engaging an agential realist critical interpretation, concepts are performative rather than descriptive or capturing of meaning – they are material enactments of the world which can be aerated to refresh and breathe life into the doing of research or inquiry.

In this entry, by diffractively reading the work of feminist philosophers such as Karen Barad, Donna Haraway and Erin Manning in relation to criticality, critique and questioning possibilities for critical interpretation in (post)qualitative research, we have responded to Latour's question on the relevance of critique and criticality and provided possibilities for adopting new theoretical frames to reinvigorate possibilities for a critical interpretative inquiry.

References

Barad, K. (2007). *Meeting the Universe Halfway: Quantum Physics and the Entanglement of Matter and Meaning.* Duke University Press.

Barad, K. (2014). Diffracting diffraction: Cutting together-apart. *Parallax,* 20 (3), 168–187, https://doi.org/10.1080/13534645.2014.927623.

Barad, K., & Gandorfer, D. (2021). Political desirings: Yearnings for mattering (,) differently. *Theory & Event,* 24 (1), 14–66. muse.jhu.edu/article/780766.

Haraway, D. (1988). Situated knowledges: The science question in feminism and the privilege of partial perspective. *Feminist Studies,* 14 (3), Autumn, 575–599.

Haraway, D. (2016). *Staying With the Trouble: Making Kin in the Chthulucene.* Durham and London: Duke University Press.

Latour, B. (2004). Why has critique run out of steam? From matters of fact to matters of concern. *Critical Inquiry,* Winter.

Manning, E. (2020). *Toward a Pragmatics of the Useless.* Duke University Press.

6 Deconstruction

Teresa K. Aslanian

Deconstruction

Deconstruction denotes a reading strategy expounded by French philosopher Jacques Derrida (1930–2004) and further developed by, among others, Julia Kristeva, which aims at destabilizing traditional modes of interpretation. The concept eludes definition, as it involves a continual critique of the assumed relationship between human and the world. As an interpretation strategy, it challenges ontological assumptions, disallows conclusive readings, and offers possibilities to explore unexamined and contradictory currents of thought at play in a text.

To understand deconstruction, we can look briefly at two ideas that influenced Derrida's thinking: Ferdinand de Saussure's (1857–1913) structural understanding of language and Martin Heidegger's (1889–1976) concept of 'Destruktion'. Heidegger (1962/1927) argued that understanding always fails because interpretation relies on underlying concepts that act as unexamined truths about existence. Thinking with these unexamined truths cannot produce an understanding of a phenomenon but only produce an understanding consisting of those preexisting ideas. For Heidegger, the destruction of these underlying concepts was necessary to truly understand phenomena. Heidegger's (1962/1927, p. 197, par. 42) work was aimed at 'arriving at the right *ontological* foundations' for understanding. Derrida adopted the idea of the impossibility of understanding phenomena due to reliance on underlying 'truths' serving as a foundation for any interpretation but abandoned the idea of any possible 'right' ontological foundation that could produce a 'true' interpretation.

Drawing this critique of metaphysics into the underlying concepts of Saussure's structuralism, Derrida (1973, 1981) questioned the unspoken but implicit idea of presence structuralism relied on. Derrida was concerned with the idea that language expresses or refers to truth (logocentrism), and its intended or unintended use to justify oppressive ideas and uphold binary thinking. Arguing that Western tradition relies on hierarchical thinking wherein one part is privileged over the other, the idea

DOI: 10.4324/9781003585046-7

of deconstruction was a way of thinking beyond traditional concepts of interpretation that relied on the idea of presence and repeatability (Derrida, 1973). To 'explode' binary thinking, Derrida (1973) insisted that the interpretation of meaning is the result of differences 'at play' in a text, an unstable interplay of shared ideas and contexts. Thus, meaning and concepts do not exist beyond the 'text'. This idea, among others, is encapsulated in his much-repeated statement, 'There is nothing outside the text' (Derrida, 1997, p. 158). Rather than denoting the meaninglessness of text, this statement conveys possibilities for liberation of meaning and for possibilities of difference.

Doing Deconstruction

Derrida (1991, p. 273) insisted deconstruction was not a method and described deconstruction as a generative way of thinking. Derrida's own work focused on beloved and influential texts, not to delegitimize their value but to expand or explode his understanding of them and our culture through an intricate exploration of the texts, seeking out points of internal contradictions that reveal unintended and underlying assumptions. Deconstruction cannot be pinned down to a specific method that can be repeated, because unrepeatability is the state of existence that deconstruction aims to reveal. It is an interpretation strategy that is sensitive to context, time, history, and the absence of repeatability. Deconstruction can be applied to the analysis of interpretations of any phenomenon that appears or registers to our senses.

A deconstructive reading can reveal assumptions, values, and systems that underlie political, social, and philosophical movements and thus can be helpful when working to weaken oppressive power structures (see Butler, 1990, 1997). A deconstructive stance has been taken to 'pick-apart' concepts the field of education relies on to expose taken-for-granted truths that perpetuate economic, social, and gender inequalities in education and early childhood education (see Canella, 1997). As a disorienting lens, deconstruction affords new perspectives on taken-for-granted ideas that are 'built into' our language and organizations. Gert Biesta (2009a, 2009b) utilizes the concept in another way, aligning himself with Derrida's (2021) insistence on deconstruction being something that doesn't need to be done because it is something that already always *occurs*. The question Biesta concerns himself with is '*how* and *when* does deconstruction occur in education?' Biesta's own act of 'witnessing' deconstruction yielded, for example, an examination of 'the subject' as a taken-for-granted idea in education that opens the possibility of the free individual, while it simultaneously delineates the idea of what it *means* to be human, excluding those who may not fit the unarticulated but still propagated description (Biesta, 2009b). Further, Biesta (2009a) grasps the idea that linguistic structures have no center to emphasize the necessity of *risk* as part of education, which ensures the limitless interpretation and variations of knowledge, rather than the unthinkable, a mechanical process of inserting a stable and delineated 'piece' of knowledge into a learner.

Deconstruction has been further developed by one of Derrida's protégés Catherine Malabou (2010, 2011), whose concept of 'plasticity' and a 'plastic reading' expands the application of deconstruction (Crockett, 2018). Malabou builds on the idea of an absent center that generates untapped possibilities 'sleeping' within a text. A plastic reading emphasizes Derrida's call to tend to 'something more' in any text. In my work (Aslanian, 2018), I generated a plastic reading to explore what *else* could be found in early childhood educators' written stories about how they experienced love in professional practice that I hadn't noticed in a prior analysis. I highlighted words in the text that seemed more 'pregnant with meaning' than other words with which they were connected. I presented the resulting list of words as a poem in which unsettling physical closeness, sensuality, and intimacy became dominant aspects of love in educational settings that are generally unspoken because of cultural associations of sexuality and sexual abuse. The sensuality and intimacy present in the original texts were masked within the language and made apparent through a plastic, deconstructive reading. Deconstruction as an interpretive method aims to explode and explore a text to generate new possibilities (Malabou & Williams, 2022). Deconstruction offers both a strategy to critically analyze otherwise taken-for-granted concepts that underlie a practice and a strategy to explore internal contradictions, revealing underlying presumptions within a text or phenomenon. Deconstruction was not conceived to inflict new violence but to '*explode* a semantic horizon' (Derrida, 1981, p. 45) of presumptions that constrict understanding and potentially oppress certain groups while privileging others. The idea is important for interpretive researchers, as it challenges and accounts for ontological presuppositions and concerns in the unintended use of text to convey mastery over others through the proliferation of ideas about what is real. A challenge to traditional methods of interpretation, its aim is not to understand what is but to excavate and explore the intertwined roots of text to generate new possibilities for the impossible.

References

Aslanian, T. K. (2018). Embracing uncertainty: A diffractive approach to love in the context of early childhood education and care. *International Journal of Early Years Education*, *26*(2), 173–185. https://doi.org/10.1080/09669760.2018.1458604.

Biesta, G. (2009a). Chapter 1: Deconstruction, justice, and the vocation of education. *Counterpoints, 323,* 15–37. http://www.jstor.org/stable/42980421.

Biesta, G. (2009b). Chapter 5: Education after deconstruction: Between event and invention. *Counterpoints, 323,* 97–113. http://www.jstor.org/stable/42980425.

Butler, J. (1990). *Gender Trouble: Feminism and the Subversion of Identity*. Routledge. http://autof.files.wordpress.com/2010/02/butler-judith-gender-trouble-feminism-and-the-subversion-of-identity–1990.pdf.

Butler, J. (1997). *Excitable Speech: A Politics of the Performative*. Routledge.

Canella, G. S. (1997). *Rethinking Early Childhood Education: Social Justice and Revolution* (Vol. 2). Peter Lang.

Crockett, C. (2018). Deconstructive plasticity: Malabou's biological materialism. In *Derrida after the End of Writing* (pp. 109–120). Fordham University Press. https://doi.org/10.2307/j.ctt1xhr741.10.

Derrida, J. (1973). *Speech and Phenomena: And Other Essays on Husserl's Theory of Signs*. Northwestern University Press.

Derrida, J. (1981). *Positions* (A. Bass, Trans.). University of Chicago Press.

Derrida, J. (1991). Letter to a Japanese Friend. In P. Kamuf (Ed.), *A Derrida Reader: Between the Blinds* (pp. 270–276). New York: Columbia University Press.

Derrida, J. (1997). *On Grammatology*. John Hopkins University Press. http://www.mohamedrabeea.com/books/book1_3997.pdf.

Derrida, J. (2021). *Deconstruction in a Nutshell: A Conversation with Jacques Derrida, With a New Introduction*. Fordham University Press.

Heidegger, M. (1962/1927). *Being and Time*. Basil Blackwell.

Malabou, C. (2010). *Plasticity at the Dusk of Writing: Dialectic, Destruction, Deconstruction*. Columbia University Press.

Malabou, C. (2011). Like a sleeping animal-philosophy between presence and absence. *Inaesthetics, 2,* Animality.

Malabou, C., & Williams, T. W. (2022). *Plasticity: The Promise of Explosion*. Edinburgh University Press.

7 Diffraction

Vivienne Bozalek and Michalinos Zembylas

Reflexivity, Diffraction, and Interpretation

Diffractive methodologies were developed in response to dissatisfaction with practices of "reflexivity", which are grounded in a representational logic and concerned more with epistemological than ontological aspects of research. Work on "reflexivity" has over the years become predominant in qualitative research methodology literature. However, there is still important conceptual work to be done putting these two practices – reflexivity and diffraction – in conversation with each other and exploring their continuities and breaks with each other as they both pertain to *interpretation*. Interpretation, the quest for the meaning of a phenomenon, has long been seen to be the heart of qualitative research (Gabriel, 2018). This entry suggests that the practices of diffraction and reflection might be understood in relation to interpretation and that doing so has important methodological implications for post-qualitative research.

Dissatisfied with epistemological practices of reflexivity grounded in representationalism, scholars immersed in feminist science studies began in the 1990s to raise questions about the theoretical assumptions and consequences of reflexivity. Donna Haraway, a pioneer in these debates, proposed the notion of *diffraction* as an alternative to reflexivity. As she expressed her misgivings at the time, "Reflexivity has been much recommended as a critical practice, but my suspicion is that reflexivity, like reflection, only displaces the same elsewhere, setting up worries about copy and original and the search for the authentic and really real" (Haraway, 1997, p. 16). Haraway suggested instead that "[w]hat we need is to make a difference in material-semiotic apparatuses, to *diffract* the rays of technoscience so that we get more promising interference patterns on the recording films of our lives and bodies" (Haraway, 1997, p. 16, added emphasis). Thus, diffraction for Haraway was suggested as a *metaphor* and a strategy for making a difference in the world. Diffraction breaks with self-reflection and its epistemological grounding, which she

DOI: 10.4324/9781003585046-8

regarded as problematic as it lures us into a reductionist way of thinking about things and words.

The notion of diffraction has been taken forward in the work of Karen Barad (2007, 2014; Barad & Gandorfer, 2021), who does not only regard diffraction as an optical metaphor but also as a *method* and a *practice* that pays attention to material engagement with data and the "relations of difference and how they matter" (Barad, 2007, p. 71). Diffraction is understood by both Barad and Haraway as a process of being attentive to how differences get made and what the effects of these differences are. Barad's notion of diffraction is derived from the physical phenomenon of diffraction in quantum physics, which they extend to other forms of knowledge production. Diffraction, from their perspective, can be used to acknowledge the influential role of the knower in knowledge production and particularly how we learn about "material configurations of the world's becoming" (Barad, 2007, p. 91). For Barad, diffraction is a useful tool highlighting the entanglement of material-discursive phenomena in the world. Diffraction is thus predicated on a *relational ontology*, an ongoing process in which matter and meaning are co-constituted.

There have already been scholarly efforts so far to theorize what reflexivity as interpretation might mean for qualitative research (e.g., Pillow, 2015) and, perhaps more obliquely, what diffractive methodologies would imply for interpretation in post-qualitative research (Barad & Gandorfer, 2021). It is, however, important to avoid setting up a binary between reflection and diffraction in relation to interpretation. If we want to be fair to the theoretical and methodological developments that have been made over the years, we might need to acknowledge that the "entanglement" of reflexivity and diffraction is one that includes continuities and breaks rather than a "story" of one versus the other. Wanda Pillow's (2015) differentiation of reflexivity as interpretation and reflexivity as genealogy, and Barad's notion of sense-making are fruitful to put into conversation with each other when attempting to reconfigure interpretation. Reflexivity as an interpretation that interrogates power/privilege in research relationships and examines who or what is benefiting from research remains important in our view. Reflexivity as genealogy has many resonances with Barad's sense-making, as it incorporates both epistemology and ontology, queers notions of progress and identity, thinks otherwise, and uses deconstruction techniques. Pillow's notion of reflexivity as genealogy relocates reflexivity from a subject-centered process to reflexivity as subjectless. Barad's sense-making is a process of getting underneath thought; asking iterative questions about what constitutes something and "thinking thinking otherwise" (Barad & Gandorfer, 2021, p. 17) are ways of rehabilitating the notion of interpretation and using it in novel ways as a form of post-qualitative inquiry.

Reflexivity and diffraction assist then in reimagining the notion of interpretation in post-qualitative research. Thinking with the theoretical impetus of notions such as reflexivity as genealogy and sense-making allows post-qualitative researchers to overcome the "ruins" left by interpretative discourses and research practices. Identifying concepts that do not fall into the trap of a binary opposition construction between interpretation or reflexivity, on the one hand, and diffraction or genealogy, on the other hand, is crucial for inventing flexible, thoughtful, and sensible research practices.

Although diffraction and reflection differ both as methodologies and as practices, being grounded in different ontologies, epistemologies, and ethics, there is some continuity in relation to certain forms of interpretation. Reflection and diffraction change interpretation in the sense that they enrich the ethical, ontological, and epistemological possibilities available when studying a phenomenon using post-qualitative inquiry. Reflection and diffraction as theoretical tools in relation to interpretation have the potential to productively question and make visible the ethico-onto-epistemological investments of research practices. A return to ethical, ontological, and epistemological questions surrounding interpretation – through tools such as diffraction – is useful to consider the present terrain of post-qualitative research.

References

Barad, K. (2007). *Meeting the Universe Halfway: Quantum Physics and the Entanglement of Matter and Meaning*. Duke University Press.

Barad, K. (2014). Diffracting diffraction: Cutting together-apart. *Parallax*, 20(3), 168–187.

Barad, K., & Gandorfer, D. (2021). Political desirings: Yearnings for mattering (,) differently. *Theory & Event*, 24(1), 14–66.

Gabriel, Y. (2018). Interpretation, reflexivity and imagination in qualitative research. In M. Ciesielska & D. Jemielniak (Eds.), *Qualitative Methodologies in Organization Studies: Volume 1 Theories and New Approaches* (pp. 137–157). Springer.

Haraway, D. (1997). *Modest_Witness@Second_Millenium: FemaleMan_Meets_OncoMouse: Feminism and Technoscience*. Routledge.

Pillow, W. (2015). Reflexivity as interpretation and genealogy in research. *Cultural Studies <-> Critical Methodologies*, 15(6), 419–434.

8 Diffractive Analysis

Kathryn Strom and Shakhnoza Kayumova

Diffractive Analysis and Interpretation

Diffraction, a classical physics term, refers to a phenomenon that occurs when a wave (e.g., sound and light) encounters an obstacle and spreads or bends around the corners as it passes by or through it. Although, according to classical physics, particles and waves were fundamentally different phenomena, when an obstacle experiment was performed with electrons at the quantum levels, scientists observed that particles exhibited wave-like characteristics. These findings in quantum physics suggest that "reality" is not objective but is entangled with our observations and measuring tools (i.e., agential cuts; Barad, 2007). In posthumanities, this quantum understanding of diffraction has been taken up to postulate that *knowing* is a material practice.

As a methodology, diffraction is informed by the assumptions of quantum physics described above and the entanglement of matter and meaning (Barad, 2014). These assumptions emerge at the juncture of feminist theories and science and technology studies (STS), also known as (feminist) new materialisms. Generally, (feminist) new materialisms shift away from the "huMan of reason" and associated disconnected, "either/or," white-male-human-supremacist, universal-transcendent, objective, ahistorical ways of thinking and being (or onto-epistemologies) that emphasize sameness or uniformity (Strom et al., 2019). New materialisms' connected methodological critique takes on the very purpose of traditional qualitative research, which is to represent lived experiences/phenomena, and the underlying belief that it is possible to "reflect" reality. Instead, new materialisms emphasize continua of subject-object, nature-culture, and material-discursive entanglements that co-constitute, or *co-make*, each other (Barad, 2007; Haraway, 2016). As such, a major feature of (feminist) new materialisms is difference and hybridity, and a foundational assumption is that of the nonrepresentational and complex nature of life. However, qualitative research has few analytic methods that can investigate phenomena in ways that map complexity, difference,

DOI: 10.4324/9781003585046-9

and hybridity. As a response, Haraway (2004) first suggested that the notion of *diffraction* might be a productive idea for researchers. She defined diffraction as a "mapping of interference" (p. 70) or the "production of difference patterns" (p. 243) that do not focus on the differences themselves as much as they do the effects of those differences. Barad (2007) moved this idea into methodology: "diffraction involves reading insights through one another in ways that help illuminate differences as they emerge: how different differences get made, what gets excluded, and how those exclusions matter" (p. 30). The diffractive methodology also has a clear set of ethico-political commitments: they are better able to attend to the material-discursive co-constitution of phenomena than traditional methodologies, demanding for researchers to politically locate (Braidotti, 2013; Haraway, 2016) or account for themselves and their "agential cuts" in these entanglements and the ways those entanglements produce knowledge (Barad, 2007/2014).

As a methodological orientation, then, the theoretical ideas and goals of diffraction are quite clear. What this looks like translated into actual methods and analytic processes, however, varies widely (and should, since diffraction is fundamentally about complexity, difference, hybridity, and emergence from local conditions). Diffractive analysis and interpretation are not about a fixed set of steps, but rather, as Davies (2014, p. 734) puts it, "[open] the possibility of seeing how something different comes to matter, not only in the world that we observe but also in our research practice." Below, we offer a brief review with situated examples of diffractive analysis and interpretive methods.

Some diffractive studies have adopted a process in which researchers "think-with" data through and with particular theoretical concepts (Fullagar & Taylor, 2022). As an example, Wolfe and Rasmussen (2020) think with affect and spacetimemattering (Barad, 2007) in their study that, through a series of vignettes, illustrates the ways the Australian school uniform dress becomes entangled with gendered bodies, producing/affecting them in different ways (i.e., creating diffraction patterns). Other studies, such as Hetherington et al. (2019), use a hybrid methodology informed by the posthuman concept of assemblage. Drawing on mixed methods, including a questionnaire, teacher interviews, photos, and focal "glow moment" data points (MacLure, 2013) in their inquiry engaging youth in the problem of ocean pollution, the researchers formed data assemblages serving as their primary unit of analysis. These "are formed and re-formed as the analysis proceeds, questioning and responding and questioning as new methodological cuts are made, new knowledge created, and new questions generated" (Hetherington et al., 2019, p. 6). Several studies also describe reading data (often from different sources and in different formats and modalities) "through" each other to produce interpretive patterns of difference. Lenz Taguchi and Palmer (2013), for example, describe reading visual and text-based data sources (photos and stories) through each other as "multiple performative agents" (p. 673). Albin Clark (2021), examining the intra-actions of early childhood teachers' documentation practices, read the material, embodied, and discursive data sources (journals, photos, audio recordings, and documentations) through each other. Others include theoretical or philosophical concepts as agents in readings, such as Invinson and Renold (2020), who read observations, field notes, and philosophical concepts from dance/movement through each other in their work with teenage girls in postindustrial Wales.

In these diffractive readings, researchers attend to the intra-actions or entanglements between data sources, thus decentering the human (Moxnes & Osgood, 2018) and attuning to both material and discursive aspects of data (Kayumova & Buxton, 2021). Hayes and colleagues, for instance, revisited their field notes from previous ethnographic studies of schools' reproduction of inequity, rewriting them to attend to entangled human and nonhuman agencies (including themselves as researchers). In mapping "sticky stories," Moxnes and Osgood (2018) focused on how "matter comes to matter" in early childhood preparation with student teachers learning to critically reflect on their practices.

Finally, diffractive analytic and interpretive practices, while not centering the human, do deliberately include the researcher as part of the analysis process – researchers co-produce the phenomena under study as well as the diffractive reading (Lenz Taguchi, 2010). Several researchers account for their own embodied attunements through a method of "glowing data" (MacLure, 2013), or hot spots, fragments of data that haunt the researcher and affect them in some way (Albin-Clark; Hetherington et al., 2019; Moxnes & Osgood, 2018; Lemieux & Thériault, 2021). This accounting for the agency of the researcher in the knowledge-construction process, including the selection of focal data based on embodied resonance, is an important part of diffraction's ethical commitments, as well as highlights its relational nature. As Lenz Taguchi (2012, p. 274) emphasizes, "I want to read with the data, understanding it as a constitutive force, working with and upon me in the event of reading it."

Although critiqued by some because it lacks a concrete set of methods, we find diffractive analysis and interpretation to be distinct from a merely theoretical approach and argue that it represents more than a mere repurposing of physics vocabulary, as shown through examples above. These diffractive methods enable us to examine how boundaries of differences are produced in our knowledge-making practices and be ethically aware that "knowing [does] not come from standing at a distance and representing but rather from a direct material engagement with the world" (Barad, 2007, p. 49).

References

Albin-Clark, J. (2021). What is documentation doing? Early childhood education teachers shifting from and between the meanings and actions of documentation practices. *Contemporary Issues in Early Childhood, 22*(2), 140–155.

Barad, K. (2007). *Meeting the Universe Halfway: Quantum Physics and the Entanglement of Matter and Meaning*. Duke University Press.

Barad, K. (2014). Diffracting diffraction: Cutting together-apart. *Parallax, 20*(3), 168–187.

Braidotti, R. (2013). *The Posthuman*. Polity Press.

Davies, B. (2014). Reading anger in early childhood intra-actions: A diffractive analysis. *Qualitative Inquiry, 20*(6), 734–741.

Fullagar, S., & Taylor, C. A. (2022). Assemblage. In K. Murris (Ed.), *A Glossary for Doing Postqualitative, New Materialist and Critical Posthumanist Research Across Disciplines* (pp. 32–33). Routledge.

Haraway, D. (2004). *The Haraway Reader*. Psychology Press.

Haraway, D. J. (2016). *Staying With the Trouble: Making Kin in the Chthulucene*. Duke University Press.

Hetherington, L., Chappell, K., Keene, H. R., & Wren, H. (2019). Creative pedagogy and environmental responsibility: A diffractive analysis of an intra-active science arts practice. In P. Burnard & L. Colluci-Gray (Eds.), *Why Science and Art Creativities Matter* (pp. 271–299). Brill.

Ivinson, G., & Renold, E. (2020). Moving with the folds of time and place: Exploring gut reactions in speculative transdisciplinary research with teen girls' in a post-industrial community. In C. Taylor, C. Hughes, & J. Ulmer (Eds.), *Transdisciplinary Feminist Research* (pp. 168–183). Routledge.

Kayumova, S., & Buxton, C. (2021). Teacher subjectivities and multiplicities of enactment: Agential realism and the case of science teacher learning and practice with multilingual Latinx students. *Professional Development in Education, 47*, 463–477.

Lemieux, A., & Thériault, V. (2021). W(e)aving: When maker literacies and diffraction collide. *Studies in the Education of Adults, 53*(2), 255–271.

Lenz Taguchi, H. L. (2010). Rethinking pedagogical practices in early childhood education: A multidimensional approach to learning and inclusion. In *Contemporary Perspectives on Early Childhood Education* (pp. 14–32). Open University Press.

Lenz Taguchi, H. L. (2012). A diffractive and Deleuzian approach to analysing interview data. *Feminist Theory, 13*(3), 265–281.

Lenz Taguchi, H., & Palmer, A. (2013). A more 'livable'school? A diffractive analysis of the performative enactments of girls' ill-/well-being with (in) school environments. *Gender and Education, 25*(6), 671–687.

MacLure, M. (2013). The wonder of data. *Cultural Studies Critical Methodologies, 13*(4), 228–232. https://doi.org/10.1177/1532708613487863.

Moxnes, A. R., & Osgood, J. (2018). Sticky stories from the classroom: From reflection to diffraction in early childhood teacher education. *Contemporary Issues in Early Childhood, 19*(3), 297–309.

Strom, K., Ringrose, J., Osgood, J., & Renold, E. (2019). PhEmaterialism: Response-able research & pedagogy. *Reconceptualizing Educational Research Methodology, 10*(2–3).

Wolfe, M. J., & Rasmussen, M. L. (2020). The affective matter of (Australian) school uniforms: The school-dress that is and does. In B. Dernikos, N. Lesko, S. McCall, & A. Niccolini (Eds.), *Mapping the Affective Turn in Education* (pp. 179–193). Routledge.

9 Digitalization

Christina MacRae, Ruth Boycott-Garnett, and Ruth Churchill Dower

Film-Making as a Virtual Rendition of Real Time

Filmmaking has been used as a uniquely scientific method foundational to many Western research disciplines. For example, it can be traced to its pivotal role in the nascent field of early childhood research in the mid-twentieth century conducted by Arnold Gesell (Curtis, 2011; MacRae, 2019; Ossmer, 2020). The authority of analogue film lies in its supposed value-free and instant-by-instant rendering of the actual. Capturing images directly onto the film in so-called real time, it becomes a research method extracting slices of time to render research subjects (in this case, children) "tangible" (Curtis, 2011). Subjects of scientific scrutiny can then be contained, coded, "and classified in a way that ephemeral behaviour cannot" (Stalpaert, 2017, p. 377). In Gesell's case, the interpretive method sought to identify universal patterns in the physiology and biological growth of the young child (Gesell, 1935). More recent research approaches to interpreting film include content analysis, thematic analysis, and narrative analysis. However, interpretation is never politically innocent, and it has habitually attributed intentions and motivations to the doings of research subjects. This bestowal of meaning can be recognised as a

> revelatory impulse that can be traced [ ] with a will to grasp and to understand the Other; "the West moistens everything with meaning, like an authoritarian religion which imposes baptism on entire peoples".
>
> (Barthes, quoted by Trinh, 1991, p. 50)

Foregrounding the colonising effects of scientific film's revelatory impulse exposes a desire for ever "more extensive and fine-grained knowledge" of research subjects (Rouse, 2005, p. 99). It is a reminder of how the materiality of analogue filmmaking has been complicit in constructing our inherited knowledge and epistemologies, shored up through research practices that construct research subjects through the act of interpreting visual data.

DOI: 10.4324/9781003585046-10

The Virtual and Material of the Digital as a Counterweight to the Extraction of Static Data

Digital media in research have often unproblematically continued to "partake in the history of scientific film" (de Freitas, 2016, p. 553). Although a film's authority has been materially implicated in its capacity to represent a transparent and visible actuality, there are other ways to approach the virtual and material dimensions of filmmaking. Digital video differs from analogue because its images are electronically rendered as binary code. While this may bolster the authenticity of analogue as a more objective record of the real, it simultaneously foregrounds the virtual and material qualities of filmmaking as such. For example, when opacity and ephemerality are welcomed, research can admit how much cannot be articulated and/or is beyond representation. A range of strategies is offered in the "Special Issue on Videography and Decolonizing Childhood" (Murris & Menning, 2019), where researchers have deployed video in ways that experimentally resist interpretative mastery.

As a counterweight to the mining of static data, Manning (2012) proposes that the virtual qualities of moving images can be used instead to materialise the incipience of the *relation* as opposed to abstracting and centring human subjects. Manning draws attention to the *relation* that is produced by bodies in co-motion. Digital media can play a role in challenging interpretative distance and what Haraway calls our siting/sighting practices (1988, p. 595) that produce bounded subjects and objects. The specific virtual qualities of video can materialise a relational, provisional, shifting approach to interpretation that unfolds both during the filmmaking process and in the retrospective replayings of this digital content. Here, the act of interpretation is a form of engagement where research becomes a process "trying to forge connections between worlds where time may be flowing at different rates, where space can have different relative scales, where we can move backward and forward in virtual time" (Lemke, 2007, p. 47). In so doing, the digital medium can undo linear notions of time and produce new social, cultural, and material relationships between researchers and sites of research.

Living in the Virtual World: Experiencing the Field Through Video Conferencing Apps

Social distancing imposed because of the COVID pandemic resulted in video conferencing being used pragmatically as new sites of fieldwork and research interaction. The Zoom video frame further troubles the possibility of "thick descriptions" (Geertz, 1973) to account for the textured, situated culture as experienced. The "real-life feel" of place and bodies is further flattened by the 2D format of the fixed frame that denies peripheral vision, absorbing indirect information such as small gestures often caught in the corner of the eye. In video calls, researchers are also constantly made aware of their own presence through seeing their name or even image on the screen. The Zoom window becomes "an active participant" interfering with what can be seen, done, and what can take place in this liminal space (Boycott-Garnett, 2023). It activates a levelling out of positionalities held between researcher and participant, limiting the degree to which the researcher can extract fixed meanings. The video call has a grainy and jerky quality: the researcher does not have access to the same touched and sensed landscape as those participating in the research, and when things are "too close to be seen properly", we are confronted with the "*unknowability* of the other" (Marks, 2000, p. xviii). Yet, simultaneously, the movement of images on the screen allows a dialogue to emerge through the visual in response to motion and sound. For example, when using dance as a research practice working with young children during the pandemic, Churchill Dower (2022) found that contact dance via Zoom, a medium that denied physical contact, was able to offer conditions that facilitated shared sensation and affect. These conditions were produced in response to movement across virtual spaces, creating a rich reciprocity of invitations and acceptances and opening the possibility of interpretation taking place both in and out of time and at subconscious levels registered across bodies. In Angell's research, instead of analysing verbal transcriptions of parent/teacher conversations conducted via Zoom, she discovered that by retrospectively replaying the digital images and responding to them through her drawing practice, she could attend to the non-vocal presences beyond words as expressed through gesture, movement, and directionality of gaze (2022).

While the example of the Zoom window affords a very particular research method, at the same time, it affirms how qualities of digital media have the potential to actively participate in more awkward, emergent, speculative, and relational interpretive practices. The digital collapses and brings together multiple sites and times: in the unfolding event of filming, in the repeated video re-replays after the event, and in the subsequent re-rendering of the digital into text or other media. This means there is no single field site/sight that can be demarcated for singular and objective interpretation. Yet, in all these multiple sites/sights and times, interpretation is ongoing and materialising the glancing connections, moments of attunement, and relationships that take form in the in-between of time and place.

References

Angell, M.-L. (2022). Drawing Attention to the Wonders of Parental Care: A Methodological Inquiry. *Qualitative Inquiry*. https://doi.org/10.1177/10778004221142807.

Boycott-Garnett, R. (2023). Zooming with Babies: Troubling a Shared Present. *Journal of Childhood Studies*, 126–132. https://doi.org/10.18357/jcs202320501.

Churchill Dower, R. (2022). Too Much Kin in the Game? The Intimate Reciprocities Available in Not Speaking. In *Cultural and Pedagogical Inquiry*, edited by A. Abdi & C. DePass. University of Alberta.

Curtis, S. (2011). "Tangible as Tissue": Arnold Gesell, Infant Behavior, and Film Analysis. *Science in Context*, *24*(3), 417–442. https://doi.org/10.1017/S0269889711000172.

de Freitas, E. (2016). The Moving Image in Education Research: Reassembling the Body in Classroom Video Data. *International Journal of Qualitative Studies in Education*, *29*(4), 553–572. https://doi.org/10.1080/09518398.2015.1077402.

Geertz, C. (1973). Thick Description: Toward an Interpretive Theory of Culture. In *The Interpretation of Cultures: Selected Essays*, edited by C. Geertz, 3–30. Basic Books.

Gesell, A. (1935). Cinemanalysis: A Method of Behavior Study. *Journal of Genetic Psychology*, *47*(1), 3–16.

Haraway, D. (1988). Situated Knowledges: The Science Question in Feminism and the Privilege of Partial Perspective. *Feminist Studies*, *14*(3), 575–599.

Lemke, J. (2007). Video Epistemology In-and Outside the Box: Traversing Attentional Spaces. In *Video Research in the Learning Sciences*, edited by R. Goldman, R. Pea, B. Barron, & S. Derry, 39–51. Mahwah, NJ: Erlbaum.

MacRae, C. (2019). Grace Taking Form. *Video Journal of Education and Pedagogy*, *4*(1), 151–166. https://doi.org/10.1163/23644583-00401003.

Manning, E. (2012). *Relationscapes. Movement, Art, Philosophy.* Cambridge: MIT Press.

Marks, L. U. (2000). *The Skin of the Film: Intercultural Cinema, Embodiment, and the Senses.* Duke University Press. https://doi.org/10.1215/9780822381372.

Murris, K., & Menning, S. F. (2019). Introduction to the Special Issue. *Video Journal of Education and Pedagogy*, *4*(1), 1–8. https://doi.org/10.1163/23644583-00401010.

Ossmer, C. (2020). Normal Development: The Photographic Dome and the Children of the Yale Psycho-Clinic. *Isis*, *111*(3), 515–541.

Rouse, J. (2005). Power/Knowledge. In *Cambridge Companion to Foucault*, edited by G. Gutting. Cambridge University Press.

Stalpaert, C. (2017). Re-Enacting Modernist Time: William Kentridge's the Refusal of Time. In *The Oxford Handbook of Dance and Reenactment*, edited by M. Franko, 375–396. Oxford University Press.

Trinh, T. Minh-ha. (1991). *When the Moon Waxes Red.* Routledge.

10 History

Svend Brinkmann

Historical Roots of Interpretation

Interpretation is often singled out as absolutely essential to the practice of qualitative inquiry. The *Cambridge Dictionary* tells us that interpretation is "an explanation or opinion of what something means". Interpretation is thus something one does in order to arrive at the meaning of something. Most scholars will argue that there is something like a figure-ground relation inherent in any interpretation of meaning. For example, the meaning of a specific word is understood on the background of the sentence in which it occurs. And the meaning of the sentence is found in the background of the larger communicative context in which it is uttered. The exclamation "I do" can mean many things, but when interpreted in the background of a wedding ceremony, it comes to have a specific meaning that is evident to everyone who is familiar with this practice and the tradition of marriage.

The point is that nothing means anything in itself, but only in the background of something else. In the broadest possible sense, interpretation thus refers to the act of relating something in the foreground to a proper background that enables one to understand the given something better. A significant part of the theoretical discussions in qualitative research has been about the validity of interpretations: What makes an interpretation convincing, trustworthy, enlightening, and strong? The notion of valid interpretation has been unfolded by several broad traditions of qualitative inquiry. This entry points to three traditions that highlight the centrality of experience, text, and action for interpretation and are represented in the first case by the phenomenological and hermeneutical traditions, in the second case by structuralism and post-structuralism, and in the third case by varieties of pragmatism.

Experience and Interpretation

Phenomenology originated with Edmund Husserl around 1900 and set itself the task of describing the essential structures of human experience

DOI: 10.4324/9781003585046-11

(Zahavi, 2003), which – most will agree – involves some kind of interpretation. It literally means the study of "phenomena", that is, of the ways in which objects and events appear to us. The goal of phenomenology is to provide descriptions of how we experience the world prior to the theories that we may formulate about it. Depression, for example, is an experienced phenomenon prior to the theories we develop about it (neurochemical, psychoanalytical, or whatever). Phenomenologists are interested in understanding the essence of phenomena, so the goal is not to describe how Jack and Jill experience depression but rather to interpret the essential properties of the experience of depression. A key interpretative procedure in this regard is what Husserl called free variation in one's imagination of the kind of phenomenon that one studies. That which one cannot remove from the phenomenon without turning the phenomenon into something else is its eidos or essence.

Hermeneutics simply is the art or practice of interpretation, and philosophers from Wilhelm Dilthey to Hans-Georg Gadamer conceived of life as an ongoing process of interpretation. As interpreting beings, we are constrained by history, culture, and what Gadamer called our prejudices, which means that there is no experience that is entirely free from interpretation (Gadamer, 1960). In a sense, then, understanding something is an act, and hermeneutic writers argue that all such understanding is to be thought of as interpretation.

Text and Interpretation

Unlike phenomenology and hermeneutics, which originated in Germany, structuralism and post-structuralism came from France. These traditions focus on interpretation as having to do with the relationship between text and context. Something means what it does only within a context, and the relationship between text and context is therefore crucial to interpretation. Structuralism began as a theory about language, developed by the Swiss linguist Ferdinand de Saussure, who argued that language is a system all of whose terms are interdependent and in which the value of one term arises only from the simultaneous presence of the others. Interpretation is about understanding the meaning of interdependent parts within a structure of language, society, or the mind.

Post-structuralists such as Jacques Derrida corrected structuralism by arguing that this larger whole is itself unstable and perspectival. This means that there simply are no fixed meanings; meaning is forever deferred in endless chains of signs that refer to other signs without ever arriving at the true, authentic meaning (Sarup, 1993). Deconstruction as an interpretive practice can be seen as a way of reading textual material (this can also be performances, organizations, and other cultural phenomena), showing how meanings move, and possibly involves suggestions and invitations into other ways that meaning may appear.

Action and Interpretation

A third central concept in relation to interpretation is action. In the pragmatist tradition, interpretation does not primarily concern figure-ground relationships in human experience or text-context relations but is about actions and consequences – doings and happenings. Something is interpreted by looking at the role it plays in human doings – actions and practices. Interpretation is performative (Guignon, 1991).

The contribution of the classical pragmatists (Peirce, James, and Dewey) can be summarized as follows: (1) Knowing is a human activity of coping with the world. Knowing is related to action, as is the interpretative relationship between what we do and what subsequently happens. (2) Knowing is something we do in our everyday lives as much as it is something that takes place in universities and laboratories. (3) Theories are therefore tools that demonstrate their validity in practice as aids in problem-solving. (4) The process of inquiry is an interpretative process that succeeds when an indeterminate situation is transformed into a more determinate situation that leads us to enriched forms of human living. Qualitative research is simply a condensed and reflective form of such interpretative inquiry.

Conclusions

Many different concepts have influenced qualitative researchers' approaches to interpretation, but this entry has highlighted just three: experience, text, and action. My goal has not been to argue that one of these approaches is correct while the others are false but rather to paint a broad picture of the variegated field of qualitative research and articulate how the important notion of interpretation has been approached by different influential historical traditions.

References

Gadamer, H. G. (1960). *Truth and Method*. Second Revised Edition Published 2000. Continuum.

Guignon, C. B. (1991). Pragmatism or hermeneutics: Epistemology after foundationalism. In D. R. Hiley, J. F. Bohman, & R. Shusterman (Eds.), *The Interpretive Turn* (pp. 81–101). Cornell University Press.

Sarup, M. (1993). *An Introductory Guide to Post-Structuralism and Postmodernism*. Second Edition. University of Georgia Press.

Zahavi, D. (2003). *Husserl's Phenomenology*. Stanford University Press.

11 Languages

Barbara Dennis, Alycia Elfreich, and Darcy Furlong

Languages of Interpretation in Qualitative Data Analysis

Interpretation, like many social science concepts, is a contested landscape. Positivist understandings of "interpretation" predominated the social sciences in the first half of the twentieth century. Through this view, interpretation was conceptualized as the link between the social scientist and the phenomena of interest. To *really know* the phenomena of interest, social scientists would want to minimize their subjective interpretations of the phenomena, producing instead as accurate a description as possible. From this perspective, interpretation is primarily a matter of accurately representing the world as it is – prior to and independent of the research going on about it. Accordingly, one's most accurate and valid interpretation would involve very little inference. Critiques of positivism (see Bentz & Shapiro, 1998, for a summary) arose from within both the natural (Barad, 2007; Haraway, 1988; Popper, 2005; Kuhn, 1970; Gadamer, 2013) and the social sciences (Geertz, 1973; Habermas, 1985; Dilthey, 2010; Weber, 1978). A thoughtful critique of positivism led to a plethora of conversations among researchers on the philosophic foundations, empirical applications, ethical practices, and social impacts of qualitative research. Ethnomethodology emerged in the 1950s as a way of looking quite closely at interactions by interpreting their meaning for participants (Garfinkel, 2016). By the 1960s, Schultz was writing about the social construction of reality, following which a variety of qualitative approaches reflect what came to be called "the interpretive turn in the social sciences" (Rabinow & Sullivan, 1987). Hermeneutics, ethnomethodology, symbolic interactionism, critical theory, and anthropological theories as represented by Geertz (1973) are examples of types of qualitative research associated with this "interpretive turn."

This interpretive turn pointed out problems with the epistemological assumptions of positivism and heightened the connection between social science and ordinary human experiences of understanding/

DOI: 10.4324/9781003585046-12

interpreting one another. In ordinary life, one persistently makes sense of their experiences, surroundings, friends, encounters, and so on through interpretive engagements. However, researchers must be more careful, deliberate, and able to warrant the interpretive claims made in association with the analysis of the data than one is expected to do in ordinary contexts.

The interpretive turn has resulted in the development of various methodological contrasts used to describe how "interpretation" is linked with "analysis." The following contrasts are meant to be illustrative and should not be considered exhaustive. Introducing them leads to this important point: The various terms and phrases associated with interpretation in qualitative data analysis are not merely neutral sets of descriptors. They imply commitments and positionalities, which will ultimately manifest as analytic decisions.

There is a set of concepts that specifically help researchers describe analytic interpretive positions relative to the interpretive perspectives of participants, such as the contrast between emic and etic. The term "emic" is used to indicate interpretations that arise from within the circle of understandings shared by participants, while "etic" indicates that the interpretations analytically apply interpretive concepts external to or outside participants' own intuitive knowledge and interpretive frameworks. Within the traditions of taking emic orientations toward interpretation, we have the language of important terms and phrases like "thick description" (Geertz) that draw our attention to the interactive contexts or situatedness of interpreting language and action. Grasping the situatedness from a participatory perspective helps researchers generate emically oriented interpretations and establish the researcher as someone able to take relative (never totalizing) insider perspectives. "Emic," "emergent," and "grounded" all refer to the practice of articulating interpretations that are plausible for the participants and internal to the data. The terms "emic" and "etic" compel qualitative researchers to locate themselves within the fields of meaning they might share intersubjectively with their participants and what might be the practical boundaries of that sharing (Carspecken, 1995). By locating one's interpretive positionality relative to the data, researchers are establishing their respective analytic perspectives.

In addition to the language that positions the researchers' interpretive orientations, some qualitative researchers use the language of logic, such as induction and deduction, to describe how interpretations might be generated in the data analysis process. Unfortunately, these terms have been commonly used to reduce our understanding of the differences between quantitative and qualitative inquiries – aligning quantitative research with deduction and qualitative research with induction. For both qualitative and quantitative inquiries, induction and deduction are relevant forms of logical inference. Imagine generating interpretations across a qualitative dataset by articulating participants' implicit notions of gender across a variety of observed actions (induction) while also applying a postcolonial feminist lens to the data (deduction) as a way to bring particular perspectives to the interpretation process. There is a necessary dialectic between induction and deduction, and they are by no means mutually exclusive.

Another contrast in the ways qualitative researchers describe the interpretive work of data analysis is to use the language of representation and reconstruction, which indicates how researchers conceptualize *meaning*. Following Foucault (2005), representation is often thought of as a social act with consequences, rather than as a neutral correspondence between symbol, meaning, and thing. Contemporary qualitative researchers explore the effects of representation by engaging with the data through a focus on what the representations do and who they benefit (Abu-Lughod, 2008; Clifford & Marcus, 1986). In this way, interpretation related to analysis is oriented toward describing systems of effects (power) that constrain and problematize assumptions of meaning. Accordingly, one must be outside the studied system of representation in order to avoid being ensnared by that same system of power (see Kuntz). This conception of power marks it as pervasively and inescapably entangled with any system of knowing. Contrastingly, reconstruction involves articulating implicit assumptions, notions of truth, and meaning (Carspecken, 2007). When taking a reconstructive approach, researchers would articulate various implicit assumptions and layers of inference through the reconstruction of meaning. One aspect of the meaning would involve exploring the effects of various interpretations and how those interpretations are intertwined with social systems, ideologies, and structures that help to contextualize ongoing social understandings and meaning (see Dennis). Both post-representational and reconstructive orientations refuse a simplistic interpretive acceptance of meaning in their analyses.

Some qualitative researchers distinguish coding from interpretation, thinking of coding as closely connected with the data and interpretation as an understanding of the relationships across codes. Consequently, some qualitative researchers, such as phenomenologists, do not actively consider their work interpretive. Such researchers might use the word "descriptive" rather than "interpretive" to characterize what they are doing. Though certainly interpretations in analyses are layered, even coding involves some form of interpretation that ought to be clarified by the researcher when explaining her analysis.

Some interpretations require a very low level of inferencing on the part of the researcher, while others will involve higher degrees of inference. A low-inference interpretation is something akin to paraphrasing, while a high-inference interpretation draws out more implicit meaning. Inferencing is always involved in interpretation because there is no way to access meaning/understanding directly. As such, all interpretations are more or less risky and have varying vulnerabilities, and all meaning and doing is uncertain (in both research and ordinary contexts).

These contrasts are not mere descriptors of how researchers conceptualize the link between interpretation and analysis – they embody commitments. Commitments can and should be articulated – whether these are theoretical, ethical, or onto-epistemological principles, for example. Generally, researchers do not swap sets of principles in and out according to the project they are engaged with. Instead, researchers' commitments can be reflected upon, deepened, thoughtfully abandoned, and so forth, but it is not really possible to analyze data interpretively without any such commitments. Making such commitments as explicit as possible strengthens qualitative research. Theoretical orientations, using labels such as "post-humanism" or "feminism," coincide with methodological commitments in the interpretive process. However, these distinctions are also muddled in generative ways – blurred boundaries, intersections, and silences bring grace to the mix. The key point is that interpretation is always a positioned and commitment-rich engagement with data that has consequences for analysis. These positions, commitments, and consequences must be reflected upon in ongoing and dialogic ways in doing qualitative research.

References

Abu-Lughod, L. (2008). Writing against culture. In T. Oakes & P. Price (Eds.), *The Cultural Geography Reader* (pp. 50–60). Routledge.

Barad, K. (2007). *Meeting the Universe Halfway: Quantum Physics and the Entanglement of Matter and Meaning*. Duke University Press.

Bentz, V. M., & Shapiro, J. J. (1998). *Mindful Inquiry in Social Research*. Sage.

Carspecken, P. F. (1995). *Critical Ethnography in Educational Research: A Theoretical and Practical Guide*. Routledge.

Carspecken, P. F. (2007). Reconstructive analyses. In G. Ritzer (Ed.), *The Blackwell Encyclopedia of Sociology*. Blackwell Publishing.

Clifford, J., & Marcus, G. E. (Eds.). (1986). *Writing Culture: The Poetics and Politics of Ethnography: A School of American Research Advanced Seminar*. University of California Press.

Dilthey, W. (2010). *Wilhelm Dilthey: Selected Works, Volume IV: Hermeneutics and the Study of History* (Vol. 4). Princeton University Press.

Foucault, M. (2005). *The Order of Things*. Routledge.

Gadamer, H. G. (2013). *Truth and Method*. A&C Black.

Garfinkel, H. (2016). Studies in ethnomethodology. In *Social Theory Re-Wired* (pp. 85–95). Routledge.

Geertz, C. (1973). *The Interpretation of Cultures* (Vol. 5019). Basic Books.
Habermas, J. (1985). *The Theory of Communicative Action, Volume 1: Reason and the Rationalization of Society*. MIT Press.
Haraway, D. (1988). Situated knowledge: The science question in feminism and the privilege of partial perspectives. *Feminist Studies, 14*(3), 575–599.
Kuhn, T. S. (1970). *The Structure of Scientific Revolutions* (Vol. 111). University of Chicago Press.
Popper, K. (2005). *The Logic of Scientific Discovery*. Routledge.
Rabinow, P., & Sullivan, W. M. (Eds.). (1987). *Interpretive Social Science: A Second Look*. University of California Press.
Weber, M. (1978). *Economy and Society: An Outline of Interpretive Sociology* (Vol. 2). University of California Press.

12 Legitimacy

Julianne Cheek

Interpretation: A Contested and Political Concept

Assumptions and claims about the legitimacy of research-derived interpretations are shaped by the set of basic beliefs (Guba & Lincoln, 1994) that we hold about what legitimate scientific knowledge is and what legitimate ways of producing that scientific knowledge are. Or put another way, "what falls within and outside the limits of legitimate inquiry" (Guba & Lincoln, 1994, p. 108). It is differences in views adopted at any point in time about what legitimate inquiry is, and why, that makes the legitimacy of qualitative interpretations a contested and political concept.

For example, we might hold the belief that subjective understandings and interpretations derived from people's experiences in context constitute "an essentially different kind of knowledge from scientifically established facts" (Crotty, 1998, p. 27). In other words, it is discovering and interpreting "facts" objectively that enables something to be deemed scientific and the research-based interpretations that produced them to be legitimate. Consequently, any research that is not designed to discover or test objective facts is deemed not legitimate and not scientific (see entry on Posts).

Such a belief effectively renders qualitative inquiry, and the interpretations derived from that inquiry, illegitimate. This is no matter how credible, authentic or trustworthy the qualitative research design and reflexive processes giving rise to those interpretations and findings are. Such a delegitimation of qualitative inquiry can result in the field of qualitative inquiry being excluded from certain research-related arenas such as the publication of, and/or funding for, research. As Kuntz reminds us, "Methodologies necessarily bring with them select assumptions about knowing and being – perspectives on the world that are heavy with political weight" (Kuntz, 2015, p. 13). This political weight gives rise to questions such as: Who can say whether interpretations based on qualitative data/inquiry are legitimate? What do they base those determinations on?

DOI: 10.4324/9781003585046-13

What mechanisms and power can they employ to ensure that their view of legitimacy is foregrounded and other possible ways of viewing what is legitimate are relegated to the margins or dismissed altogether?

Effects of This Politics of Legitimacy When Publishing Research

There are various arenas where we see the outworking of effects of perspectives on the world that are heavy with such political weight. One such arena is the publication of research. Decisions about what will and will not be published are made based on what editors, peers and publishing houses consider legitimate science, and any interpretations arising therefrom, to be.

For example, such a politics of legitimation surfaced when the editors of the *British Medical Journal* (BMJ) decided to delegitimate qualitative inquiry, declaring that qualitative inquiry was of an extremely low priority for the journal. Such extremely low priority was premised on the editors' views that qualitative studies are not interesting to clinical readers, add very little, and are of not much practical value (see Greenhalgh et al., 2016). These views effectively acted to delegitimize, and therefore exclude, any reporting of qualitative studies in this journal. It characterized qualitative research and the evidence it produces as not, or at the very least less, legitimate.

What the editors did not declare was the basis for, and justification of, these views – the perceived inadequacy of qualitative interpretations. A perception based on their undeclared assumptions and basic beliefs about what science is and is not and, therefore, what research-derived interpretations can, and cannot be, provide legitimate "evidence" on which to base practice. Undeclared assumptions that are based on a politics of evidence that challenges the perceived value and legitimacy of qualitative inquiry (see Cheek, 2018).

The Intersection(s) of Research Funding and "Legitimate" Interpretations

Research-related funding bodies also adopt views about what type of research can be considered scientific. They then use those views to base decisions on when awarding funds to researchers in competitive funding rounds. To be successful and win funding, it is important for an application for funding to be in keeping with those views. This is because it is proposals employing methods of collecting and interpreting data that the funding body deems to be worthy, valid, and therefore legitimate, which are funded. Consequently, in effect, funding bodies attempt to "regulate scientific inquiry by defining what good science is" (Denzin, 2009, p. 13) and what are legitimate scientific and research-based interpretations.

Historically, many powerful funding bodies have been hostile to, or at the least extremely sceptical of, qualitative inquiry premised on the understanding of science and evidence drawn largely from biomedical models of research (Denzin, 2009, 2010; Cheek, 2022). Qualitative inquiry has been deemed illegitimate because it cannot produce "legitimate" scientific evidence on which to base policy and practice (Cheek, 2022). This has resulted in most qualitative inquiries being disqualified and deemed not worthy of funding based on the type of knowledge that qualitative interpretations produce.

A Conservative Challenge to Legitimacy of Qualitative Interpretation

The connection between publication and funding, and perceived legitimate and illegitimate modes of inquiry, reflects what has come to be known as a conservative challenge to qualitative inquiry.

It is conservative because it reflects and acts to maintain, sustain, and produce new variants of the historic "heavy emphasis on quantification in science. . . . Scientific maturity is commonly believed to emerge as the degree of quantification found within in a given field increases" (Guba & Lincoln, 1994, pp. 105–106).

It is a challenge because such a view is premised on (often undeclared) basic beliefs about the superiority of quantification and its assumed link with objectivity. Such beliefs challenge and act to delegitimize the intellectual work of the art of interpretation (Denzin, 1989; Denzin & Lincoln, 2000) – the cornerstone of qualitative inquiry.

References

Cheek, J. (2018). The BMJ Debate and What It Tells Us About Who Says What, When and Where, About Our Qualitative Inquiry. In N. K. Denzin & M. D. Giardina (Eds.), *Qualitative Inquiry in the Public Sphere* (pp. 50–65). New York and Oxon: Routledge.

Cheek, J. (2022). The Impact of Funding on Ways Qualitative Research Is Thought About and Designed. In U. Flick (Ed.), *Handbook of Qualitative Research Design* (pp. 339–354). London; Thousand Oaks: SAGE.

Crotty, M. (1998). *The Foundations of Social Research: Meaning and Perspective in the Research Process*. New South Wales: Allen & Unwin.

Denzin, N. K. (1989). *The Research Act* (3rd ed.). New Jersey: Prentice Hall.

Denzin, N. K. (2009). *Qualitative Inquiry Under Fire*. Walnut Creek, CA: Left Coast Press.

Denzin, N. K. (2010). *The Qualitative Manifesto: A Call to Arms*. Walnut Creek, CA: Left Coast Press, Inc.

Denzin, N. K., & Lincoln, Y. S. (2000). Preface. In N. K. Denzin & Y. S. Lincoln (Eds.), *Handbook of Qualitative Research* (2nd ed., pp. ix–xx). SAGE.

Greenhalgh, T., Annandale, E., Ashcroft, R., Barlow, J., Black, N., Bleakley, A., Boaden, R., Braithwaite, J., Britten, N., Carnevale, F., Checkland, K., & Ziebland, S. (2016). An Open Letter to The BMJ Editors on Qualitative Research. *BMJ, 352*. https://doi.org/10.1136/bmj.i563.

Guba, E. G., & Lincoln, Y. S. (1994). Competing Paradigms in Qualitative Research. In N. K. Denzin & Y. S. Lincoln (Eds.), *Handbook of Qualitative Research* (pp. 105–117). Thousand Oaks, CA: SAGE.

Kuntz, A. M. (2015). *The Responsible Methodologist: Inquiry, Truth-Telling, and Social Justice*. Walnut Creek: Left Coast Press.

13 Mountain

Abhik Chakraborty

Thinking With Mountains: A Qualitative Foray on Geological Time

Qualitative research is usually preoccupied with human society and its multifariousness, but occasionally, it needs to make forays into nonhuman realms. Mountains and the geological time they embody beckon us to appreciate nonhuman agency and simultaneously challenge qualitative interpretation to embrace nonhuman materiality. There are a few notable endeavors such as Clark's (2011) portrayal of a primordial and 'inhuman' nature and Cohen's (2015) account of the lithic dimensions of 'stone.' Such interpretations are in some ways related to earlier forays: Latour's famous 'Actor-Network-Theory' (ANT) (Latour, 1999), which seeks to portray all human experiences as enmeshed in a fluid, multidirectional, incessantly changing web of things, and Hacking's (2000) evoking of the inhuman scale behind the formation of dolomite as a critique of social constructionism being prominent examples. Mountains and geological time inform us that human knowledge and worldviews form only one sliver of a vast range of possibilities incited by matter and its nonhuman/more-than-human agency, and instead of starting with specific knowledge claims or worldviews, qualitative researchers are challenged to appreciate how our interpretations are shaped by the performative agency of mountain matter.

There are already some frameworks that explore multiple relationalities of humans and nonhumans in qualitative research. Both ANT (Murdoch, 2001; Kanazawa, 2018) and the work of phenomenologist David Abram (Abram, 1997) uphold multiple interpretations of reality or phenomena. In ANT, actants and networks are mutually co-constructionist, open-ended, and evolve in an organic and fluid manner. Abram advanced the important claim that humans are corporeally *embedded* within the living biosphere. Yet, in both these approaches, humans still somehow retain the center stage. While Abram recognized that the human intellect is shaped by multiple nonhuman entities, he nevertheless

DOI: 10.4324/9781003585046-14

observed that we maintain the ability to alter or suspend our engagements with phenomena. And, despite acknowledging the agency of our nonhuman counterparts, the orbit of ANT remains essentially 'social' (Whatmore, 2002).

In contrast, Clark (2011) argues that not only are we embedded in the dynamic earth, but also there is a 'radical asymmetry' in that relationship. This is because earthly nature predates our species and is essentially *inhuman* in scale and function. Cohen (2015) seeks to separate the agency of stone as 'dense' – an attribute that is borne out of layers of time being condensed in its lithic body. He advances the concept of 'geophilia' – an affective relationship with all types of earthly matter – which essentially rejects the centrality of the human experience. These important insights help us to understand that human-more-than-human-nonhuman relationalities are not completely open-ended and that there is an *ordering* in those relationalities that stems out of the fact that we humans are *antecedent* to many natural bodies and phenomena – a necessary condition for us humans inhabiting a Cenozoic planet located along the tapestry of geological time.

This is where mountains – the most mature form of the evolving continental crust, which appears to be the key behind the successful evolution of multifarious life (Gray, 2013; Chakraborty & Gray, 2020) – offer a further key for qualitative interpretation of long-present matter and its nonhuman materiality. Mountains literally embody a temporal sequence that is inhuman in scale. With the journey of the continents colliding and separating as a consequence of the Wilson cycle of opening and closure of ocean basins (Condie, 2016), sections of ancient seafloors are eventually thrust up miles into the air to become peaks soaring into the sky, and these lofty bodies are then incised and reduced by multiple agents of erosion that work in tandem with our planet's slow dance in its cosmic orbit (Herman et al., 2021; Zalasiewicz, 2012). The forces that operate to raise and denude mountains over geological time can neither be adequately described by any human language or experience, nor are they amenable to our everyday lived experiences. What is more is that mountains are cradles of biodiversity across deep time (Rahbek et al., 2019) – implying that we are all products of mountains. Thus, mountains resist becoming the *object* of this inquiry, as our location at a specific point in the continuum of geological time forecloses our ability to completely comprehend their orogenic agency; we are always left with a *partial* view of them because of being foreordained as such in geological time. A related insight is that some part of the mountain is *already within us* when we are looking at it, and this *already within us* state is not merely a matter of perception but denotes the very sense of our being in such a fundamental way that the subject-object duality in the inquiry breaks down; not exactly in the sense phenomenologists proclaim but due to the facticity of mountains that finds the human subject as an *antecedent* to their ancestral lithic bodies. This line of interpretation resonates well with Karen Barad's (2007) observations that we are a part of the nature we seek to understand, that our nonhuman counterparts actively partake in the world's ongoing engagements toward making itself intelligible, and that humans are part of the ongoing reconfiguration of worldly phenomena. Extending these observations, we can visualize geological time as an emergent phenomenon borne out of interactions between myriad abiotic and biotic forms of matter – or perhaps more aptly, as the evolution of mountains implies, as a phenomenon that is infused with the historicity of nonhuman matter spread over scales that are only partially knowable to us. This partiality of our knowledge implies that mountains retain the capacity to shape it over a temporal scale in ways that we cannot fully comprehend, and this realization in turn requires us to forego any epistemic knowledge claim that posits us on an equal plane with geologically active matter. Perhaps it is better to be attentive to the differential representations of geologically active mountain matter (such as glaciated peaks, cirques, barren rocky outcrops, sediment flows, and underwater mountains) and explore how their different historicities result in inherently pluralistic patterns of abiotic and biotic material cohabitation. Because *uncertainty* and a *deep understanding* of multifarious phenomena are strengths of qualitative interpretation, perhaps qualitative researchers can fruitfully respond to this beckoning of intimacy with vibrant mountain matter.

Is there a conclusive note from this line of inquiry? There is hardly a necessity for one because of the fact that complexity and messiness in findings can be major strengths of qualitative research (see Law, 2004). Even some natural scientists have advocated the rejection of simplistic and clear-cut findings for complex problems (Stirling, 2010), and this foray both upholds the irreducible complexity in qualitative interpretations and resists being subsumed within a particular research paradigm.

References

Abram, D. (1997). *The spell of the sensuous: Perception and language in a more-than-human world*. Vintage Books.

Barad, K. (2007). Meeting the *universe halfway*: Quantum *physics and the entanglement of matter and meaning*. Duke University Press.

Chakraborty, A., & Gray, M. (2020). A call for mainstreaming geodiversity in nature conservation research and praxis. *Journal for Nature Conservation*, 56, 125862. https://doi.org/10.1016/j.jnc.2020.125862.

Clark, N. (2011). Inhuman nature: *Sociable* life on a dynamic planet. Sage.

Cohen, J. J. (2015). Stone: *An* ecology of the inhuman. University of Minnesota Press.
Condie, K. C. (2016). Earth as an evolving planetary system (3rd ed.). London: Academic Press.
Gray, M. (2013). Geodiversity: *Valuing* and conserving abiotic nature. Wiley-Blackwell.
Hacking, I. (2000). *The social construction of what?* Harvard University Press.
Herman, F., Doncker, F. D., Delaney, I., Prasicek, G., & Koppes, M. (2021). The impact of glaciers on mountain erosion. *Nature Reviews Earth and Environment*, 2, 422–435. https://doi.org/10.1038/s43017-021-00165-9.
Kanazawa, M. (2018). Research *methods for environmental studies: A social science approach*. Routledge.
Latour, B. (1999). On recalling ANT. In J. Law & J. Hassard (Eds.), Actor *network theory and after* (pp. 15–25). Blackwell.
Law, J. (2004). After method: *Mess* in social science research. Routledge.
Murdoch, J. (2001). Ecologising sociology: Actor-network theory, co-construction and the problem of human exemptionalism. *Sociology*, 35(1). https://doi.org/10.1177/00380385010350010.
Rahbek, C., Borregaard, M. K., Antonelli, A., Colwell, R. K., Holt, B. G., Nogues-Bravo, D., Rasmussen, C. M. Ø., Richardson, K., Rosing, M. T., Whittaker, R. J., & Fjeldså, J. (2019). Building mountain biodiversity: Geological and evolutionary processes. *Science*, 365, 1114–1119. https://doi.org/10.1126/science.aax0151.
Stirling, A. (2010). Keep it complex. Nature, 468, 1029–1031. https://doi.org/10.1038/4681029a.
Whatmore, S. (2002). Hybrid geographies: *Natures*, cultures, spaces. Sage.
Zalasiewicz, J. (2012). *The planet in a pebble*. Oxford University Press.

14 Multimodality

Denise Newfield and Sarah Crinall

Multimodality and Interpretation

Interpretation in studies of multimodality – a burgeoning field since the 1990s – is a matter of meaning, analysis and explanation, as in the humanities more generally. The activity of interpretation concerns the description, analysis and explanation of the meaning of multimodal texts and interactions through careful observation and theoretically based analysis, which follows procedures and methods developed (and still being developed) in the field. Martin Thomas (2009), for example, laments the lack of 'established procedures for the development of document corpora' in his analysis of the rhetorical relations between texts and images in toothpaste packages from the United Kingdom and Taiwan. Multimodal researchers working within humanist and representationalist perspectives focus on the multiple ways in which meaning is produced, represented and communicated in the contemporary communication landscape – through images, sounds and music, the use of space and the body, as well as through language. Multimodality as a disciplinary domain has sought to broaden the focus of description and interpretation from language and linguistic texts to systems, interactions and texts in all modes. A number of specialist concepts have been, and continue to be, developed for this purpose. Key amongst these are multimodality (the way in which different modes are utilised in human communication and representation), mode (a semiotic resource which is socially shaped and which can be realised through diverse materials and on different media platforms such as the screen or whiteboard), materiality (the material out of which meaning is made), affordance (the potential, logic or grammar of a particular mode and its concomitant constraints), sign (the bringing together of signifier and signified in a single unit of meaning for the purpose of representation and communication), semiotics (the study of semiotic resources and their uses) and semiosis (the process of sign- or meaning-making in humans). Analysis of speech might therefore consider – in addition to language – sonic modes of vocal quality, volume

DOI: 10.4324/9781003585046-15

and inflection and bodily modes of gesture and facial expression, while analysis of writing would take visual elements into account – handwriting style, font, space and colour, as well as visual images, should they be included – in addition to words. Studies in multimodality have been applied in a range of domains – semiotics, communication, education, digital media, culture, health, architecture, entertainment and art – often to reveal hidden or naturalised meanings and subtle processes. Typical interpretative tasks might concern the analysis of advertisements to show underlying ideologies, of multimodal texts produced by students in classrooms to show signs of learning, or of small pieces of conversation to show the minute and meticulous orchestration of different modes such as eye movements, gaze, gestures, vocal inflections and silences.

Challenging the notion of 'system' in traditional semiotics, multimodal social semiotics – a popular framework in the field of multimodality (Kress, 1997, 2010; Kress & van Leeuwen, 1996; Jewitt, 2014; Stein, 2008) – provides a social account of the sign and emphasises the agency and contextual situation of the sign-maker. Student texts might be analysed for signs of engagement in learning or lack of it, or for indigenous influences (Stein, 2008; Newfield, 2014), and product packaging might be analysed for its gender, class and cultural biases and values. Another framework for the study of multimodality, multimodal discourse analysis (O'Halloran, 2004), considers language and other semiotic systems as tools for creating meaning and structuring thought and reality through the ideational, interpersonal and textual metafunctions derived from linguist Halliday (1985). Jewitt's *Handbook of Multimodal Analysis* (2014) provides a toolkit for multimodal analysis and interpretation through theoretical discussions and case studies. It includes themes of technology, culture, identity and power in multimodal communication and representation and shows how multimodal analyses may draw on other disciplines including anthropology, visual studies and sociolinguistics. Jewitt, Bezemer and O'Halloran's *Introducing Multimodality* (2016) outlines the different strands of multimodal theory and research. Bateman et al. (2016) attempt to provide practical methodologies for research into communicative artefacts, materials, events and performances 'on their own merits' – and from different starting points – rather than by pushing them into categories designed for other items of investigation.

In most versions of multimodality studies, representation is seen as the way ideas, experience and reality are 'represented' not only in words but also in images, sounds and other modes or all of these in complex orchestrations. The analyst's task is to see how the representations work and what meanings are made through them. Thus, much of the focus is on texts as 'congealing' or 'fixing' of processes.

However, a focus on process presents a foundational challenge to multimodality studies away from images and text as a representation of the world. Manning's (2020) interest is in the 'in-act', the indeterminate movement towards an event. Barad (2007) preferences uncovering what lies behind, what is in excess of and what images and text *do* in the world. With indeterminacy always transient, the process comes into focus.

Process philosophies are challenging the long-time human-centred obsession with finding finite meaning, as an ethic, a politic of care beyond the human, and for love – resisting the urge to 'interrogate the world and mine it for meaning' (MacLure, 2017, p. xvii), as if we humans ever could come to fully understand it (St. Pierre, 2011). This 'interpretosis' (Deleuze & Guattari, 1987) – a human-centred neurosis – is finding scholars interrogating colonialism's dominance (Rosiek et al., 2020), researching 'for love, for care, for being with others . . . to choose how we are going to be in that story' (Rose, 2014, n.p). Indigenist ways of knowing, being and doing have understood this relationality for millennia (Graham, 1999; Martin & Mirraboopa, 2003).

One way to shake Newtonian interpretive foundations has been to disrupt rational language and expression. Structural experimentations such as Karen Barad's 'Quantum entanglements' (2010) demonstrated how images and text can pull philosophers into writing indeterminacy into the world, not writing to determine meaning (Barad & Gandorfer, 2021). Crinall's bodyplaceblogging (2019) is one example of writing using multiple modes, producing and involving expressive word/image beings indeterminately in a visual poetic set of events (Figure 14.1). They capture a relational collaboration past, and then, the author is open to the blog post altering the direction of any writing that engulfs it next (e.g. this entry). Instead of representation, the issue of expressibility and collaboration becomes central to post-qualitative production (Barad & Gandorfer, 2021):

With expression in focus, in post-qualitative approaches, there can be a shift to becoming playful with the artefacts of research (Crinall, 2019), as a commitment to staying with questions and relationality and remaining inclusive of indeterminacy and disruptive of the human/non-human divide. Where interpretation in multimodality focuses on orchestration of modes, a post-qualitative research moment might take an emergent 'line of flight' (Deleuze & Guattari, 1987), collaborating with an aspect of theory or, like here, an odd object lying on a floor during thesis writing. Is the author playing with the human/nature divide? Did the shearwater's affective agency produce the surprising aside? Importantly, a second effect is available when following lines of flight with 'glowing data' (MacLure, 2017). Post-qualitative approaches remain open to the lively matter of this artefact as an 'image/text being' in this writing moment too. For example, to glance at the blog post now, the term *stranded between* catches an eye. One might note in collaboration with the theory being

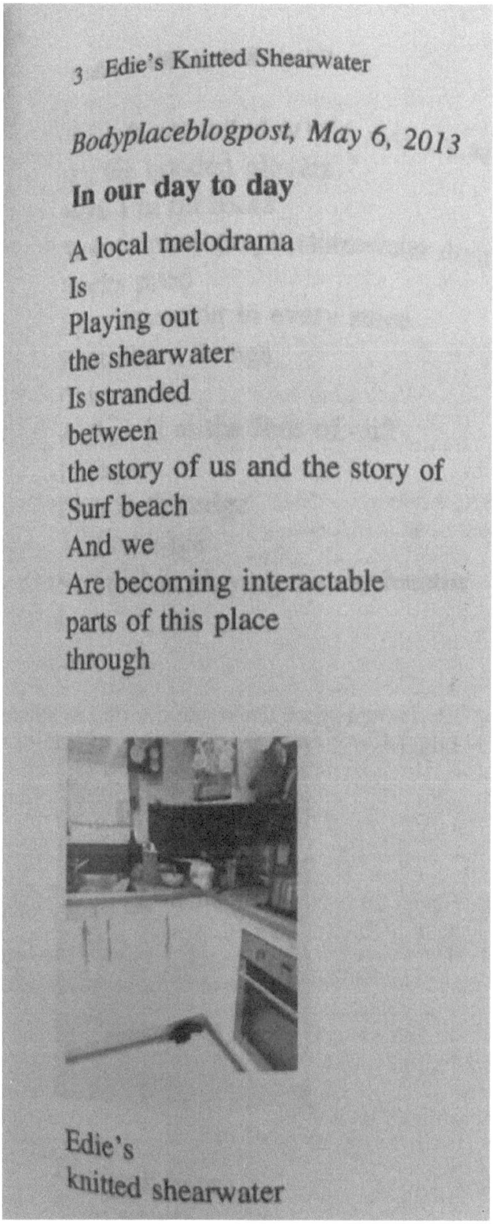
3 Edie's Knitted Shearwater

Bodyplaceblogpost, May 6, 2013

In our day to day

A local melodrama
Is
Playing out
the shearwater
Is stranded
between
the story of us and the story of
Surf beach
And we
Are becoming interactable
parts of this place
through

Edie's
knitted shearwater

Figure 14.1 'In our day to day'

Source: Crinall (2019). Chapter 3: Edie's knitted shearwater. Included with the publisher's permission.

articulated in this entry that there is not a shift in identity in this blog post. Instead, like all relational philosophies, it leaves us stranded between to revel in it without conclusion. It is an in/conclusive, in/determinate and in/finitely moving image/text mark on these pages. Moving attention between the story of humans and places as relationality calls for, and holding our conclusion here between how images and texts can produce forms and formats, is in question. The fact that we have attempted to highlight the *melodrama* of 'what is multimodality between the qualitative and post-qualitative realm?' is 'always, already enough' (Barad & Gandorfer, 2021).

References

Barad, K. (2007). *Meeting the universe halfway*. Duke University Press.
Barad, K. (2010). Quantum entanglements and hauntological relations of inheritance: Dis/continuities, spacetime enfoldings, and justice-to-come. *Derrida Today*, 3(2), 240–268.
Barad, K., & Gandorfer, D. (2021). Political desirings: Yearning for mattering (,) differently. *Theory and Event*, 24(1), 14–66.
Bateman, J., Wildfeuer, J., & Hiippala, T. (2016). *Foundations, research and analysis: A problem-oriented introduction*. De Gruyter Mouton.
Crinall, S. M. (2019). *Sustaining childhood natures: The art of becoming with water*. Springer Press.
Deleuze, G., & Guattari, F. (1987). *A thousand plateaus* (B. Massumi, Trans.). The Athlone Press.
Graham, M. (1999). Some thoughts about the philosophical underpinnings of aboriginal worldviews. *World Views Environment Culture Religion*, 3(2), 105–118. https://doi.org/10.1163/156853599X00090.
Halliday, M. A. K. (1985). *Introduction to functional grammar.* Edward Arnold.
Jewitt, C. (Ed.). (2014). *The Routledge handbook of multimodal analysis* (2nd ed.). Routledge.
Jewitt, C., Bezemer, J., & O'Halloran, K. (2016). *Introducing multimodality.* Routledge.
Kress, G. (1997). *Before writing: Rethinking the paths to literacy.* Routledge.
Kress, G. (2010). *Multimodality: A social semiotic approach to contemporary communication*. Routledge.
Kress, G., & van Leeuwen, T. (1996). *Reading images: The grammar of visual design*. Routledge.
MacLure, M. (2017). Foreward. In Mirka Koro-Ljungberg, Teija Löytönen, & Marek Tesar (Eds.), *Disrupting data in qualitative inquiry: Entanglements with the post-critical and post-anthropocentric*. Peter Lang Publishers.
Manning, E. (2020). Radical pedagogies and metamodellings of knowledge in the making. *CrisTal (Critical Studies in Teaching and Learning)*, 8(Special Issue), 1–16.

Martin, K., & Mirraboopa, B. (2003). Ways of knowing, being and doing: A theoretical framework and methods for indigenous and indigenist re-search. *Journal of Australian Studies*, 27(76), 203–214.

Newfield, D. (2014). Transformation, transduction and the transmodal moment. In C. Jewitt (Ed.), *The Routledge handbook of multimodal analysis* (2nd ed., pp. 100–114). Routledge.

O'Halloran, K. (Ed.). (2004). *Multimodal discourse analysis*. Continuum.

Rose, D. B. (2014, March 24). *Deborah bird rose: Country and the gift* [video]. YouTube. https://www.youtube.com/watch?v=suSbvoAw0g4

Rosiek, J., Snyder, J., & Pratt, S. (2020). The new materialisms and indigenous theories of non-human agency: Making the case for respectful anti-colonial engagement. *Qualitative Inquiry*, 26(3), 331–346.

Stein, P. (2008). *Multimodal pedagogies in diverse classrooms: Representation, rights and resources*. Routledge.

St. Pierre, E. A. (2011). Post qualitative research: The critique and the coming after. In N. K. Denzin & Y. S. Lincoln (Eds.), *The Sage handbook of qualitative research* (4th ed., pp. 611–625). Sage.

Thomas, M. (2009). *Localising pack messages: A framework for corpus-based cross-cultural multimodal analysis*. PhD Thesis, University of Leeds.

15 Multispecies

Tuure Tammi and Pauliina Rautio

Multispecies Studies

Situated within the broader return to materiality and critique of human exceptionalism across social sciences, humanities and arts, multispecies studies maintain that human-only stories serve poorly in the current times of biosocial destruction. Thus, the concept 'multispecies', often with emphasis on justice, signals a focus on unequal vulnerabilities between humans as well as other species and ecosystems.

The object of multispecies studies is accountability: the revision of dominant hierarchies of life and worth to enable the flourishing of all species. The understandings needed for this arise from situated encounters, which also shape inquiry. Thus, interpretation, as part of the inquiry process, is understood more as a productive matter of response and world-making than that of representation and assigning meaning in retrospect. To crumble and revise dominant hierarchies requires engagement with earthy, problematic and uncertain ways of knowing: new ways of thinking are crafted in action rather than prescribed as interpretations. While the concept of species is employed, explorations are not limited to those animals 'big like us' as traditionally in human-animal studies. Some scholars even go beyond 'living subjects' such as animals, plants, fungi and microbes and suggest that the boundary between 'animate' and 'inanimate' is fluid, and the two are regarded as existing on the same continuum. The concept of multispecies thus denotes a shift toward 'kinds and their multiplicities' to discuss new understandings, relationships and accountabilities (van Dooren et al., 2016; Kirksey & Helmreich, 2010; Ogden et al., 2013). Another effort to dismantle the firmly seated hierarchies is an insistence on attentiveness to not only those kinds cherished by the majority of humans – the fluffy, cuddly, hand-fitting, cute – but also 'uncharismatic' and 'unloved others' and those encounters producing mixed, difficult and awkward affects (Lorimer, 2014; Rose & van Dooren, 2011).

DOI: 10.4324/9781003585046-16

Multispecies studies seek to cultivate 'arts of attentiveness', often through ethnography, passionately and intimately immersing oneself in the lives of particular others (i.e., giving attention to them) while learning that these beings are also attentive themselves (i.e., things matter to them in various ways). Authors such as Vinciane Despret (2016), Donna Haraway (2003, 2008), Anna Tsing (2010) and Thom van Dooren et al. (2016) all show that vital to a multispecies approach is acknowledging that humans are not the only beings with interpretive capabilities – abilities to pay attention and respond, classify, recognize, differentiate and articulate. Thus, multispecies studies are ethically committed to seeking ways to learn how to account for these multiple and coexisting interpretations, the storied matter, in order to better respond to another for the sake of mutual flourishing while keeping in mind also the more critical questions of when kinds are shaping and being shaped by biological, cultural, economic, technological and political forces.

Multispecies Childhoods

Research of children's beliefs, attitudes, interactions or the like with other (mostly animal) species is plentiful and conducted with a wide array of interpretative methods from statistical analysis to exploratory focus group discussions to open-ended interviews. Interpretation in such studies is focused on identifying and describing multiplicities and related processes, but lacking an overall focus on uneven conditions of livability cannot be characterized as 'multispecies studies'. Research of multispecies phenomena in childhood or children's lives with a focus on shared vulnerabilities and accountability is less frequent but has been gaining attention recently. For instance, a handbook section edited by Tracy Young and Pauliina Rautio (2020) includes entries that highlight the feelings, attitudes or understandings of multiple individuals and the search for connections and accountabilities, and provides new definitions or paths of inquiry.

There is nothing particularly new about multispecies (childhood) studies per se; the understanding of human life as linked to other species has been well versed historically and with many Indigenous cultures (e.g., Pulkkinen & Lindfors, 2016). The newness arises from our current historical-scientific context: the means through which we can learn about the multispecies conditions of livability are different from the past, thus enabling new kinds of understandings of coexistence.

Multispecies childhood studies often focus on situated particularities, such as children's encounters with ants and worms (Taylor & Pacini-Ketchabaw, 2015), and, through analyzing 'minor events', proceeds to a tracing of threads that connect the events to countless other events and phenomena. The focus is on the multiplicity of engaging agents, such as when more-than-human families are approached through a focus on the processes of how companion animals participate in defining and shaping the notion of family (Irvine & Cilia, 2017). Sometimes, the focus is on speculatively producing a child-animal encounter with the help of a particular concept, such as 'childhoodnature', to find out what directions of inquiry open up (Duhn & Quinones, 2020). Multispecies childhood studies are often also geared at proliferation – finding and creating ever-new stories and frames for phenomena (Hohti & Tammi, 2019) with an additive empiricist logic (Rautio et al., 2022).

Interpretation, in multispecies childhood studies, seems to be embedded in the inquiry process on the whole and not targeted at research materials only. Thus, inquiry becomes an active doing, an experimental proposition for 'how we might attune ourselves to the challenges of our changing world in a way that is open and creative' (Rooney, 2018); it is then through a processual and embodied participation that interpretation takes place and reveals what is possible. A prerequisite of this kind of interpretation is the imperative that the complexity of life is not simplified by imposing dualist categories (human/animal, nature/culture) evidenced in a dominant 'one-world' mono-species perspective. Unsubscribing such human exceptionalism makes research more complicated and speculative but also more 'worldly' (Haraway, 2008).

Concerns and Debates

As posthumanist and new materialist work, such as multispecies studies, grounds ontogenetically on relationality and becoming, some scholars have raised concerns regarding the possibility that connection (or entanglement) is treated as good in itself rendering some troubles, such as slow processes of violence, imperceptible (e.g., Kopnina, 2017; Giraud, 2019). In short, attending to multifaceted affects and effects is critical.

On the one hand, multispecies childhood refers to an understanding that childhood is not only constituted as a sociocultural phenomenon but is also shaping and being shaped by encounters and relations with other animacies and forces. On the other hand, it denotes the urgency to figure out ways to move nonhierarchically to account for these kinds and their multiplicities. Peter Kraftl (2020), although not explicitly a multispecies scholar, suggests holding multiple foci while asking, 'where are children, exactly'. The children are included, not so much 'in their own right' as if detached individuals, but among the kinds and currents materializing in particular (everyday) circumstances. A focus on childhood in multispecies studies can offer a salient critique. For instance, children can disrupt the idea of representation through their grammatically and habitually 'incorrect' responses (MacLure, 2016), while other-than-human children may be gazing back at us in surprising places further challenging the taken-for-granted research protocols and methods (Tammi et al., 2023).

Whether at the horse stable, in the schoolbook, in an animal rescue, at the dinner table, nearby open-cutting or by a kindergarten terrarium, multispecies childhoods are lived in specific ways that are always also political – they are consequential to and have consequences for sustenance, damage, decease and recuperation of multispecies worlds. In these specific worlds, 'details of affect, materiality, labor, and ethics could be critically highlighted and interrogated even if the joys and troubles of co-living do not cease to exist simultaneously' (Rautio et al., 2021).

References

Despret, V. (2016). *What Would Animals Say If We Asked the Right Questions?* University of Minnesota Press.

Duhn, I., & Quinones, G. (2020). Eye-to-eye with otherness: A childhoodnature figuration. In A. Cutter-Mackenzie-Knowles, K. Malone, & E. Barratt (Eds.), *Research Handbook on Childhoodnature* (pp. 87–102). Springer.

Giraud, E. H. (2019). *What Comes After Entanglement?* Duke University Press.

Haraway, D. (2003). *The Companion Species Manifesto: Dogs, People and Significant Otherness*. Prickly Paradigm Press.

Haraway, D. (2008). *When Species Meet*. University of Minnesota Press.

Hohti, R., & Tammi, T. (2019). The greenhouse effect: Multispecies childhood and non-innocent relations of care. *Childhood*, 26(2), 169–185. https://doi.org/10.1177/0907568219826263.

Irvine, L., & Cilia, L. (2017). More-than-human families: Pets, people, and practices in multispecies households. *Sociology Compass*, 11(2), e12455.

Kirksey, S., & Helmreich, S. (2010). The emergence of multispecies ethnography. *Cultural Anthropology*, 25(4), 545–576.

Kopnina, H. (2017). Beyond multispecies ethnography: Engaging with violence and animal rights in anthropology. *Critique of Anthropology*, 37(3), 333–357.

Kraftl, P. (2020). *After Childhood: Re-Thinking Environment, Materiality and Media in Children's Lives*. Routledge.

Lorimer, J. (2014). On auks and awkwardness. *Environmental Humanities*, 4(1), 195–205.

MacLure, M. (2016). The refrain of the a-grammatical child: Finding another language in/for qualitative research. *Cultural Studies ↔ Critical Methodologies*, 16(2), 173–182.

Ogden, L., Hall, B., & Tanita, K. (2013). Animals, plants, people, and things: A review of multispecies ethnography. *Environment and Society*, 4(1), 5–24.

Pulkkinen, R., & Lindfors, S. (2016). *Suomalaisen kansanuskon sanakirja*. Gaudeamus.

Rautio, P., Hohti, R., Tammi, T., & Ylirisku, H. (2022). Multiple worlds and strange objects: Environmental education research as an additive practice. *Australian Journal of Environmental Education*, 38(3–4), 214–226.

Rautio, P., Tammi, T., & Hohti, R. (2021). Children after the animal turn: Child–animal relations and multispecies scholarship. In N. Yelland, L. Peters, N. Fairchild, M. Tesar, & M. S. Perez (Eds.), *The Sage Handbook of Global Childhoods* (Vol. 1, pp. 341–352). Sage.

Rooney, T. (2019). Weathering time: Walking with young children in a changing climate. *Children's Geographies*, 17(2), 177–189. https://doi.org/10.1080/14733285.2018.1474172.

Rose, D. B., & van Dooren, T. (2011). Introduction. *Australian Humanities Review*, 50(1), 1–4. https://doi.org/10.22459/AHR.50.2011.03.

Tammi, T., Hohti, R., & Rautio, P. (2023). From child–animal relations to multispecies assemblages and other-than-human childhoods. *Barn–forskning om barn og barndom i Norden*, 41(2–3), 140–156.

Taylor, A., & Pacini-Ketchabaw, V. (2015). Learning with children, ants, and worms in the Anthropocene: Towards a common world pedagogy of multispecies vulnerability. *Pedagogy, Culture & Society*, 23(4), 507–529.

Tsing, A. L. (2010). Arts of inclusion, or, how to love a mushroom. *Manoa*, 22(2), 191–203.

van Dooren, T., Kirksey, E., & Münster, U. (2016). Multispecies studies: Cultivating arts of attentiveness. *Environmental Humanities*, 8(1), 1–23.

Young, T., & Rautio, P. (2020). Childhoodnature animal relations: Section overview. In A. Cutter-Mackenzie-Knowles, K. Malone, & E. Barratt (Eds.), *Research Handbook on Childhoodnature: Assemblages of Childhood and Nature Research* (pp. 1211–1220). Springer.

16 (No) Method

Walter Omar Kohan

(No) Method (in Advance): A Freirean Perspective

This entry discusses the issue of the method itself. It is inspired by Paulo Freire, who is usually associated with a literacy method in critical and popular pedagogy. The method can be understood, broadly, as a path to interpret, research and teach. The aim of this writing is to put into question the way a researcher relates to the methodological dimension of her practice. It will show how the issue of the method is affected by epistemological and political dimensions through a childlike exercise inspired by Freire's pedagogy of the question.

In contemporary neocapitalistic times, the pressure to apply strict methods and obtain certain outcomes is highly and increasingly conditioning research practices. Nevertheless, the methodological demand upon educators and researchers has always existed as well as the philosophical and political implications of such requests. This demand crosses self-proclaimed democratic or authoritarian regimes and different times and cultural contexts, and it is precisely these implications that this text will hopefully help to think about and appraise.

In an interview with N. Pelandré, when questioned about his method, Freire focuses on what interests him most, the philosophical and political aspects of the method:

> I would begin to answer your question by making some considerations that seem to me, from an epistemological point of view, important. Considerations concerning the expression you used, and which is not only used by you, but by everyone, when you asked me about "the method". *I would prefer to say that I have no method.* What I had, when I was very young, 30 or 40 years ago, it doesn't matter the time, was curiosity on one side and political commitment, in the face of the renegades, the denied, the forbidden to read the word, rereading the world, on the other. What I tried to do, and continue doing today, was to have an understanding of educational practice that

DOI: 10.4324/9781003585046-17

> I would call critical or dialectical, within which, necessarily, there is a certain methodology, a certain method.
>
> (Pelandré, 2014, p. 14, our translation and italics)

It is worth paying attention to every detail of this answer. Let's read it slowly. First, Paulo Freire offers what seems a surprising declaration: "I would prefer to say that I have no method". Then, he justifies this statement and gives other names to what is usually called method: curiosity and political commitment to the renegades, the denied and the forbidden to read the word. These concepts are situated in different places in the educational/research process. If a method is more relevant in the middle of the process, that is inside of it, in practice, curiosity is more strongly implicated at the beginning or even before an educational/research practice; in turn, political commitment guides the educational/research journey from the end or even beyond it, like a horizon. If the method has to do with "how" we educate/research, curiosity speaks about "why" we educate/research, and political commitment refers to "for what" we do it. If curiosity is a principle from which we start and which impels us to walk, the method is a way to transit to walk, and political commitment alludes to the meaning or sense of walking. Paulo Freire considers that there is no way or method to walk before walking. The title of his conversational book with the popular US educator Myles Horton is very suggestive in this sense: *We made the road by walking* (Freire & Horton, 1990). The road (method) emerges from the exercise of walking.

Thus, Paulo Freire makes it explicit that his main concern has always been with questions of principles and meaning or sense of educating/researching, much more than with questions of paths, methods or methodologies. He had a method in the sense that it was necessary to walk in some way, but – and this clarification is also very significant – it was more a way of knowing (epistemological) than of teaching (pedagogical). Why is this clarification important for readers of this *Key Concepts* book? Because it shows that for Paulo Freire, if it was necessary to anticipate a certain way or form of knowing, it was not necessary for pedagogical reasons, to teach, but for epistemological reasons, since it was a way of producing knowledge; then, the Paulo Freire method, if any, was more epistemological than pedagogical. Researchers reading this *Key Concept* interested in Paulo Freire might pay special attention to this dimension: it touches them as knowers, interpreters, researchers and *coinventor*s of a more beautiful and fair world.

Curiosity is a form of childhood that needs to be nurtured by the educator and the educant: to educate means, above all, to nurture the curiosity that underlies every question that is the key point of a pedagogy of the question and the most problematic aspect of a pedagogy of the answer (Freire & Faundez, 2013 [1985]; Kohan, 2021a, 2021b). For researchers of any age reading this *Key Concept*, the signs are at least double: childhood is something to be kept alive in oneself at any age, and chronological children teach about be(com)ing researchers by the way they engage with the world and between themselves. On the other hand, the political commitment to the "renegades, denied, forbidden to read the word, rereading the world" is another irreplaceable and non-negotiable component of educators and researchers of all ages. It is education as politics: a politics that is not defined by a particular content but by a commitment to a political subject characterised, in our societies, by its negation.

That there is no method in advance does not mean that the educator/researcher does not need to know the things needed for a meaningful passage along this path or that there are no strategies and techniques that can help in the educational/research process. What Paulo Freire seems to be suggesting is that no technique can be self-contained, immutable or considered applicable to any context and that each educator/researcher needs to interpret and relate creatively to whatever technique she practises.

When N. Pelandré asks Freire if the guidelines were the same in the different countries where the method was applied, he affirms the importance of critically thinking about the practice and the subordination of the method to its political and philosophical dimension, and he states, "each educator is a method". Every researcher is a method. Those who educate/research cannot be separated from how they educate/research because educators/researchers are constituted and find themselves in the way they affirm their epistemological, philosophical and political commitment in an emancipatory practice. And there is no way to separate the educational/research method from the educating/researching subjectivity. The task of finding one's own way is irreplaceable for those who dedicate their lives to popular education/research because, when found through a dialogical and thoughtful process, it shows the aspirations of an educational/researching life committed to a more beautiful, fair and loving world.

References

Freire, P., & Faundez, A. (2013 [1985]). *Por una pedagogía de la pregunta. Crítica a una educación basada en respuestas a preguntas inexistentes*. Buenos Aires: Siglo XXI.

Freire, P., & Horton, M. (1990). *We Made the Road by Walking*. Temple University Press.

Kohan, W. (2021a). *Paulo Freire: A Philosophical Biography*. Bloomsbury.

Kohan, W. (2021b). *Paulo Freire: um menino de 100 anos.* NEFI. filoeduc.org/editora.

Pelandré, N. L. (2014). Entrevista com Paulo Freire, EJA EM DEBATE, Florianópolis, a. 3, n. 4, p. 13–27, jul. Entrevista publicada inicialmente. em N. L. Pelandré (Ed.), *Ensinar e Aprender com Paulo Freire: 40 horas 40 anos depois* (3. ed. rev). Florianópolis: Editora da UFSC.

17 Performance

Annette Arlander and Pilvi Lumia Porkola

Performance as Research (PAR)

Why discuss performance in the context of interpretation? Performance in the wide sense of action and production is at the core when gathering or generating research data and when sharing and spreading research results. But why link it to interpretation? Perhaps simply to remind us that interpretation is performance or production, too. One connection is the historical link in traditional forms of understanding artistic performance when performance and interpretation were almost conflated, and an interpretation of a preexisting text or score was the basis of performance. In that context, a musician performing a musical work is interpreting it. Oral interpretation was important in the development of performance studies, but the term is not used much within performance as research (PAR). Performance could be understood as an alternative to interpretation; rather than assuming a preexisting text that is interpreted, a text can be produced in or through performance. Rather than a spectator interpreting an action, the one who analyzes an action as a performance frames it as one. 'The move toward performative alternatives to representationalism shifts the focus . . . to matters of practices/doings/actions' (Barad, 2003, p. 802).

In performance studies, the term 'performance' is used and understood in a wide sense; any cultural activity can be viewed as a performance, whether a political demonstration or a football match, worship or shopping. All human activity can be considered from a performance perspective. While we can study something that 'is' performance, we can also study almost anything 'as' performance, as (if it were) a performance, an activity (Schechner, 2006, p. 38). A clear distinction between a thing that 'is' something and a thing considered 'as' something is hard to maintain and has been criticized (Kershaw & Nicholson, 2011, p. 4) and discussed (Cull, 2014).

The many uses of the term 'performance' from Richard Schechner's understanding of performances as actions or restored behavior, via

DOI: 10.4324/9781003585046-18

Marvin Carlson's (1996) point that the use of performance as a metaphor has helped focus attention from the 'what' to the 'how' of culture, to Jon McKenzie's (2001) claim that performance, besides an analytical tool, is also a disciplinary instrument ('perform, or else'), have all contributed to the development of PAR. Peggy Phelan's (1993) famous claim based on performance art – performance's only life is in the present, and performance becomes itself through disappearance – has been contested, however. Philip Auslander (1999) has deconstructed the opposition between live and mediated forms; Rebecca Schneider (2001) has maintained that archives disappear as well, while performances constitute bodily techniques of remembering, and Diana Taylor (2003) has stressed embodied practices as important systems of transmitting knowledge in her discussion of the archive and the repertory. Performance could be understood as repetition, as in Butler's (1990) performative reiterations of normative gender production or in Barad's (2003) posthumanist performativity, which concerns all material-discursive practices that produce difference.

Performance in artistic research, in pedagogy or in social sciences brings forth different aspects of the concept and enables different kinds of approaches in research. In artistic research, one can produce research material by making performances or thinking through performances. (Cull O Maoilearca, 2019). In pedagogy, the classroom situation can be analyzed through the concept of performance: how do space, duration and materiality effect the learning environment? How is teaching and learning embodied? (Perry & Medina, 2011; Porkola, 2020). In social sciences, performance can offer a methodological lens to analyze political events or demonstrations focusing on doing, on action and on process (Taylor, 2003, p. 3).

The theory-practice divide and the valorization of textual over embodied knowledge within academia have long been criticized (Conquergood, 2004). Modern arts and humanities have structured a gap between practice and analysis, while practice as research and PAR claim that creative production can constitute intellectual inquiry (Riley & Hunter, 2009, xv). Parts of performance studies known as PAR use creative practice as a methodological approach, emphasizing the interchange between doing and knowing and a close relationship between research and performance because many scholars are practicing artists, many artists theorize their way of working, and it is difficult to maintain the distinct categories of artwork and research (Borggreen & Gade, 2013, p. 12). An important influence in the development of PAR has been the Performance as Research Working Group of the International Federation for Theatre Research (IFTR), founded by Jacqueline Martin and Baz Kershaw in 2004 (see PAR website[1]), as well as the publications related to it (Arlander et al., 2018).

As a mode of 'investigating by doing', like learning by doing, PAR is a methodological approach and is increasingly used as one methodology among others within the humanities and social sciences, although most frequently in theater and performance studies, where people feel comfortable with the term 'performance as research'. In the United Kingdom, the term 'practice as research' is mostly preferred. The latter could be criticized for separating theory and practice and for not distinguishing artistic practices and other practices; theorizing is a practice, too. That criticism goes for performance as well; all research is performed in some manner. One problem with the term 'performance as research' is that in performing arts, performance is often understood as the result, not as the action or process.

Performance as research could be seen as part of the qualitative paradigm or as part of a separate performative paradigm. Taking the title from J.L. Austin's (1999) speech act theory, Brad Haseman (2006) suggested performative research as an alternative to the qualitative and quantitative paradigms and distinguished two typical characteristics: (1) the research is not problem-based but led by 'an enthusiasm of practice', and (2) the research outputs and claims to knowing are made through the symbolic language of the practice, which is the principal research activity (Haseman, 2006, p. 7). Performative research uses symbolic data, including material practices, moving and still images, sounds and music, live action and digital code, and these data work performatively; not only do they describe the research, as so-called constative utterances do, but also, like performative utterances, they do something in the world (Haseman, 2006, p. 6). The performative paradigm has remained largely unexplored with a few exceptions (Bolt, 2008, 2016). In an early text, Patricia Leavy, who has propagated arts-based methods for research in social sciences, presents a schema rather similar to Haseman's, upholding arts-based research as the third mode, though as an expansion of the qualitative paradigm (Leavy, 2009, p. 256).

A performative turn in the arts, the transformation of the artwork into a shared event, has implications for PAR as well. The conditions for the production and reception of art have transformed the artwork into an event (Fischer-Lichte, 2008, p. 28) with no difference between the artwork and its production; the audience becomes part of the work. Performance as an event that is mutually produced by all parties present brings us close to what Conquergood (2004) proposed in his time, research as something jointly performed by everybody involved.

When discussing performance as a method, Diana Taylor (2003, p. 16) states that more interesting than defining the term is to consider what it allows us to do; performance offers bodily, affective and active/performative ways of knowing alongside the written. This is a crucial dimension of PAR. The multiple uses and meanings of the term 'performance' are reflected in what can be done with performance in terms of research topic, methodology and dissemination. Performance is a practical tool for generating research material through various types of workshops as well as an effective way of sharing research outputs or challenging academic conventions for conference presentations. As a notion, performance can be more than an alternative to interpretation and provide a path to move beyond representations.

Note

1 PAR website: https://iftr.org/working-groups/performance-as-research

References

Arlander, A., Barton, B., Dreyer-Lude, M., & Spatz, B. (Eds.). (2018). *Performance as Research: Knowledge, Methods, Impact*. Routledge.

Auslander, P. (1999). *Liveness*. Routledge.

Austin, J. L. (1999). *How to Do Things With Words*. Harvard University Press.

Barad, K. (2003). Posthumanist performativity: Toward an understanding of how matter comes to matter. *Signs: Journal of Women in Culture and Society*, 28(3), 801–831.

Bolt, B. (2008). A performative paradigm for the creative arts? *Working Papers in Art and Design*, 5. ISSN 1466–4917. https://www.herts.ac.uk/__data/assets/pdf_file/0015/12417/WPIAAD_vol5_bolt.pdf.

Bolt, B. (2016). Artistic research: A performative paradigm? *PARSE Journal*, 3, 129–142.

Borggreen, G., & Gade, R. (2013). Introduction: The archive in performance studies. In Gunhild Borggreen and Rune Gade (Eds.), *Performing Archives/Archives of Performance* (pp. 9–30). Museum Tusculanum Press.

Butler, J. (1990). *Gender Trouble*. Routledge.

Carlson, M. (1996). *Performance: A Critical Instruction*. Routledge.

Conquergood, D. (2004). Performance studies, interventions and radical research. In Henry Bial (Ed.), *The Performance Studies Reader*. Routledge.

Cull, L. (2014). Performance philosophy – Staging a new field. In Laura Cull & Alice Lagay (Eds.), *Encounters in Performance Philosophy* (pp. 15–38). Palgrave Macmillan.

Cull O Maoilearca, L. (2019). Artistic research and performance. In Paolo de Assis & Lucia D'Errico (Eds.), *Artistic Research: Charting a Field in Expansion* (pp. 146–174). Rowman & Littlefield.

Fischer-Lichte, E. (2008). *Transformative Power of Performance: A New Aesthetics*. Routledge.

Haseman, B. (2006). A manifesto for performative research. *Media International Australia Incorporating Culture and Policy: Quarterly Journal of Media Research and Resources*, 118, 98–106.
Kershaw, B., & Nicholson, H. (Eds.). (2011). *Research Methods in Theatre and Performance*. Edinburgh University Press.
Leavy, P. (2009). *Method Meets Art – Arts-Based Research Practice*. The Guilford Press.
Mckenzie, J. (2001). *Perform or Else – From Discipline to Performance*. Routledge.
Perry, M., & Medina, C. A. (2011). Embodiment and performance in pedagogy research: Investigating the possibility of the body in curriculum experience. *Journal of Curriculum Theorizing*, 27(3), 62–75.
Phelan, P. (1993). *Unmarked. The Politics of Performance*. Routledge.
Porkola, P. (2020). Miten tila ja aika merkityksellistyvät? Uusmaterialistia ja esitystutkimuksellisia näkökulmia pedagogiikkaan. In Heli Kauppila & Kai Lehikoinen (Eds.), *Toiminnasta sanoiksi. Puheenvuoroja oman työn kehittämisestä taidealojen yliopistopedagogisessa koulutuksessa*. Taideyliopisto.
Riley, S. R., & Hunter, L. (Eds.). (2009). *Mapping Landscapes for Performance as Research – Scholarly Acts and Creative Cartographies*. Palgrave Macmillan.
Schechner, R. (2006). *Performance Studies: An Introduction* (2nd ed.). Routledge.
Schneider, R. (2001). Archives – Performance remains. *Performance Research*, 6(2), 100–108.
Taylor, D. (2003). *The Archive and the Repertoire: Performing Cultural Memories in the Americas*. Duke University Press.

18 Politics

Juha Suoranta

Politics of Interpretation in Times of War

Social realities are distinguished by their varying intensities, but war is beyond all other intensities since war contains the most destructive elements of human endeavor. It simultaneously colonizes all spheres of life, the whole interpretative framework of the human mind, and wipes other topics, such as climate change, from the international agenda.

War links the semiosphere with the biosphere, and words become flesh. The instantaneous nature of the semiosphere shapes perceptions and dominates the view of war in the digital age. As C. Wright Mills noted in the late 1950s, "The media not only give us information; they guide our very experiences. Our standards of credulity, our standards of reality, tend to be set by these media rather than by our own fragmentary experience" (Mills, 2000, p. 311). Nonetheless, the essential fact remains that war involves ordinary people killing one another in the name of vicious, ideologically charged national symbols manipulated by thoughtless leaders.

However, in the horrendous conditions of war and its interpretations, the fragmentary experience Mills writes about highlights the catastrophe of war from an individual's point of view. Thus, the importance of using the first person singular in the politics of interpretation; the following passage is from my notebook:

> I recall the early morning of February 24th, 2022. I wake up at my usual time, around 6:00 a.m. It is my birthday, a perfect day for a small party. As I checked the news headlines, it struck me that this would not be a day of celebration but a fateful day for Europe and the world. Russia's authoritarian leader Vladimir Putin brutally invaded Ukraine an hour earlier, at 5 am, thus expanding his ongoing war in east Ukraine since 2014. A few days earlier, Putin had bafflingly declared that Ukraine has turned against Russia, although "[s]ince time immemorial, the people living in the south-west of what has

DOI: 10.4324/9781003585046-19

historically been Russian land have called themselves Russians and Orthodox Christians" (Putin, 2022).

What can ordinary social scientists do on a morning like that? They put their television on and search for more information. They try to make sense, understand, and even explain – even if they feel helpless in the face of atrocities. They clean their desks of not-so-urgent tasks. Many world-class intellectuals, such as Chomsky, McLaren (2022), and Žižek (2022a, 2022b, 2022c), have done so to grasp the horrible situation. They attempted to demonstrate how the Ukraine war is linked not only to Europe but also to world history, the legacy of the Cold War, and international power politics. Wood (2022) returned to the past, believing it may teach something: "It is hard not to see parallels here with the twilight of the Belle Epoque. Then, as now, inter-imperial tensions fed a headlong arms race. Then, as now, too, public opinion readily rallied behind national governments." These and other troubadours of truth offered invaluable political and structural insights to interpret and understand the logic of the Ukraine war. However, the most fundamental point remained: the war destroyed people's lives and homes, the natural environment, and any efforts to build a sustainable future.

What, then, are social scientists' options for the politics of interpretations in times of war? What do they need to reconsider when interpreting it situated in a time and politics of war?

First, they must make sure not to join forces with warmongers or crackpot realists who, in C. Wright Mills' description, are at the top of the world they possess. As Mills wrote:

> [They] have constructed a paranoid reality all their own; in the name of practicality they have projected a Utopian image of capitalism. They have replaced the responsible interpretation of events with the disguise of events by a maze of public relations; respect for public debate with unshrewd notions of psychological warfare; intellectual ability with agility of the sound, mediocre judgment; the capacity to elaborate alternatives and gauge their consequences with the executive stance
>
> (Mills, 1956/2000, p. 356)

Therefore, social scientists must carefully analyze media-mediated interpretations of war from all sides and witness and document war conditions, casualties, collateral damage, and people's everyday lives in dire circumstances in the war zone and elsewhere.

Second, scholars may use their public voice and communicate on all fronts. They can contact local social scientists who live and work in the war zones to support them and obtain direct information. Likewise, they can connect with social movements and aid organizations inside and outside the conflict area. In addition, they can participate in writing and speaking publicly, teaching their students, debating, marching, demonstrating, and choosing other active roles in the conflict.

Third, social scientists need to acknowledge the formation of the current "macrostructural unconscious," that is, to understand how ideologies constitute our everyday lives and how that understanding "enables us to reflect upon why we acquiesce to the root-and-branch deceptions of the corporate media, who possess all the latest technologies for manufacturing our loyalty to them and the capitalist class which they serve" (McLaren, 2022). Scholars should also revisit an existential question about their work: their task as textual workers. What does it mean to submit texts in scientific journals in safety compared to living in conditions where surviving and dying are part of everyday life?

"If we as intellectuals do not define and re-define reality, who will?" wondered C. Wright Mills (2008, p. 221) decades ago. During the 1950s, he sought to expand the social sciences to encompass a global and comparative study of international relations and the critical possibilities of peace. He hoped that the new scope could help social scientists and intellectuals stand against the numbing and repressing ideologies of the two superpowers of the time, the United States and the Soviet Union. Mills aimed at viable democratic societies where people, and not only leaders, have a responsible say in local and global issues in terms of world peace (Mills, 1958).

In its brutal and inhumane manner, the fact of war suggests that the world has become an ever more fluid place consisting of uncertain, unpredictable, and chaotic flows and events (Law, 2004, p. 3). As Law puts it, "if we want to think about the messes of reality at all then we're going to have to teach ourselves to think, to practice, to relate, and to know in new ways." In war conditions, we must develop a *nouveau* art of interpretation. Thus, we must learn to know the many realities of the world:

> through the hungers, tastes, discomforts, or pains of our bodies. These would be forms of knowing as embodiment. Perhaps we will need to know them through "private" emotions that open us to worlds of sensibilities, passions, intuitions, fears and betrayals
>
> (Law, 2004, pp. 2–3)

As a result, we might realize that social scientists can respond to the drums of war by expanding our interpretive horizon to Sartre's singular universal (Crittenden, 1998). This means that, as researchers of the social world, we cannot apprehend individual experience without its relations to the history and the whole social fabric in which a knowing/feeling subject is embodied.

References

Crittenden, P. (1998). The Singular Universal in Jean-Paul Sartre, *Literature and Aesthetics*, *8*, 29–42. https://openjournals.library.sydney.edu.au/LA/article/view/5230.

Law, J. (2004). *After Method: Mess in Social Science Research*. Routledge.

McLaren, P. (2022). Reflections on the Performative Dimensions of the War in Ukraine, *PESA Agora*. https://pesaagora.com/columns/reflections-on-the-performative-dimensions-of-the-war-in-ukraine/. Retrieved April 19, 2022.

Mills, C. W. (1958). *The Causes of the World War Three*. Simon and Schuster.

Mills, C. W. (2000). *The Power Elite*. Oxford University Press.

Mills, C. W. (2008). The Decline of the Left. In Mills, C. W., Ed., *The Politics of Truth: Selected Writings of C. Wright Mills* (pp. 213–222). Oxford University Press.

Putin, V. (2022). Address by the President of the Russian Federation, February 21, 2022. http://en.kremlin.ru/events/president/news/67828. Retrieved April 20, 2022.

Wood, T. (2022). Matrix of War. *New Left Review 133/134*. https://newleftreview.org/issues/ii133/articles/tony-wood-matrix-of-war

Žižek, S. (2022a). What Does Defending Europe Mean? *Project Syndicate*, March 2, 2022. https://www.project-syndicate.org/onpoint/hot-peace-putins-war-as-clash-of-civilization-by-slavoj-zizek-2022–03?referral=a8de3b. Retrieved April 19, 2022.

Žižek, S. (2022b). From Cold War to Hot Peace, *Project Syndicate*, March 18, 2022. https://www.project-syndicate.org/onpoint/hot-peace-putins-war-as-clash-of-civilization-by-slavoj-zizek-2022–03?referral=a8de3b. Retrieved April 19, 2022.

Žižek, S. (2022c). War in the World That Stands for Nothing, *Project Syndicate*, April 18, 2022. https://www.project-syndicate.org/commentary/russia-ukraine-war-highlights-truths-about-global-capitalism-by-slavoj-zizek-2022–04?referral=a8de3b. Retrieved April 19, 2022.

19 Posts

Mirka Koro and Ananí M. Vasquez

Posts Shaping Interpretation

Interpretation has been considered an essential element of qualitative inquiry throughout its history. Theoretically, interpretive qualitative inquiry has often been situated in phenomenologies, hermeneutics, and constructivisms. However, today's theoretical extensions and practical applications are very different from their 1970s and 1980s counterparts. Onto-epistemological and ethical shifts have opened interpretation, as a concept and practice, to new directions. Furthermore, post-qualitative thinking and doing have also shaped the ways in which interpretation in qualitative inquiry is currently being used, practiced, and written about.

Some of the earlier onto-epistemological assumptions of interpretation supported the singularity of the researcher and the separation of interpretive research activities from the knowers (e.g., participants). The need to learn about others' experiences, thoughts, feelings, and social practices and to assign them significance (Willig, 2014) possibly originated from the beliefs that all things are inherently separate and therefore unknown to others. Interpretation was enacted as a knowledge production practice that extracted data from communities and participants' experiences and then moved these forms of information to different academic settings. Other onto-epistemological assumptions of past interpretation practices supported co-constructed and multiple realities or truths, which led researchers to focus on shared understanding and co-constructed meanings. In this way, interpretation itself was understood to be culturally and socially constructed (Grbich, 2013). Furthermore, whatever the positionality of the researcher, interpretation as analytic thought that includes reflection on complexity and that goes beyond description (Saldaña, 2015) is considered generalizable from local to universal contexts, enabling the researcher to apply this understanding to other individuals and situations. At the same time, it has been acknowledged that even when research is collaboratively constructed between researchers and participants, there is still the danger of abuse through uneven power relations (Willig, 2014).

DOI: 10.4324/9781003585046-20

Uneven power relations and politics in research have raised many ethical concerns. For example, interpretation has been used as a technique for learning about different cultures and peoples, sometimes by the colonizers themselves. Some qualitative researchers justified entering communities as outsiders by citing ethics of empathy and care. As "voyagers" collecting foreign knowledge, scholars attempted to represent the unknown voices and experiences of others. However, despite good intentions, researchers sometimes produced overgeneralizations and oversimplifications based on their cultural assumptions and political biases. Occasionally, interpretation was used as a strategy to justify the representation of otherness through the potentially damaging perspective of colonial and hegemonic discourses.

These issues, among others, have mobilized some researchers to shift to alternative onto-epistemological assumptions, moving from interpretation as extractive, knowledge producing, and meaning-making toward relational, material, and affective processes. For many qualitative researchers coming from critical and post-traditions, interpretation as a form of relating has a more material, experimental, and anti-anthropocentric emphasis. One could argue that many qualitative scholars still engage in potentially interpretive practices and work through various levels of understanding with data/materials/lives. However, the processes and aims of these practices have shifted. Theoretical movements associated with the posts (postmodernism, post-structuralism, postcolonialism, posthumanism, etc.) enable contemporary scholars to critique and work against desires and demands to interpret and understand the inaccessible other. Post critiques ask difficult and complex onto-epistemological questions about the role of theories, the nature of understanding, the benefits of interpretation, and the beneficiaries of interpretive practices (see Grant, 2014; Anker & Felski, 2017).

Some critical scholars (e.g., Denzin & Lincoln, 2018; Koro-Ljungberg, 2016; Povinelli, 2016; Somekh & Schwandt, 2007) keep in mind that the act of interpretation is influenced by continuously changing cultural elements and values and that interpretative practices are impacted by cultural structural changes, global trends, and theoretical movements. For example, the global environmental crisis has stimulated more research that decenters humans and emphasizes the role of material things, post- and more-than-human elements of world-building. These decenterings and reorientations have also shaped the knowledge, becomings, and worldings that interpretations create. No longer is interpretation an entirely human activity of forced or potentially artificial understanding of the foreign; neither is it ethically appropriate nor is it necessarily desirable. Rather, some scholars working through interpretation draw from post-intentional phenomenologies (Vagle, 2014) and phenomenologies of practice (Van Manen, 2014), where interpretation is connected to everyday practices of living. In addition, interpretation as a desire for cultural insider-ness and access to the unknown is replaced by relationality within situational knowledge and through nearness of (ontological) differences.

A material, ethical, and ontological relationality has become a preferred way among qualitative researchers to interact inside and outside continuously shifting research contexts. For example, Henry (cited in Van Manen, 2014) brought materialism to phenomenological interpretation and focused on immanence, the material beginnings of meanings. Henry's material experiences demonstrated interest in the original source of material and of consciousness and the deepest roots of affects. Life feels itself; we feel life itself within us when it is happening, and these forces constitute interpretations. Similarly, for some scholars, methods "become a practice of being *inside* a research event" and of "becoming entangled in relations" (Springgay & Truman, 2017). Serres (2008), in turn, approached interpretation through sense phenomenology and sensuous perceptions of the world. Serres utilized wordplay, paintings, poetry, and mythologies, among others, to generate interpretations, which are beyond ordinary language and language's knowledgeability.

In these examples, relationality disrupts data extraction through the issue of representation, which has often been the precursor to and product of interpretation (as meaning-making). A "crisis" of representation through relationality can be described as a blurring of science and literature and thresholding toward "multivocality, contested meanings, paradigmatic controversies, and new textual forms" (Lincoln et al., 2018, p. 145), which also leads to productive responses to ethical questions of interpretation. Relationality in qualitative inquiry leads to "the reality that human life is connected to and dependent on other species and the land" and that implies that the research not only documents the relationality of life but also "influences these conditions in small or significant ways; it thus implies action and increased accountability to people and place" (Tuck & McKenzie, 2015, p. 4).

Issues of representation and ethics have also changed the way scholars write about interpretation and their interpretive practices. For example, according to Van Manen (2014), writing interpretatively belongs to the unrepresentable. Authors are encouraged to bring to the present the writing that escapes all representations, and this kind of interpretation can be rather challenging because "language substitutes itself for the phenomenon that it tries to describe" (Van Manen, 2014, p. 370). From this perspective, words kill the objects of presentation. Darkness is the method (of writing and interpretation) for Van Manen, which problematizes visibility and obviousness in writing and interpretation. Alternately, Bachelard (2014) brought poetics to writing processes, interpretation, and phenomenology. From this viewpoint, language functions as an image and text, and both need to be transcended. Mediation and scholars' sensitization to multiple variations within interpretative processes are needed for well-rounded interpretation. This sensitizing may also enable us to intimately meet ourselves from within, the poetic forms of interpretation (and phenomenology) being interpretations of the soul. Bachelard reminds us that by creating and remembering interpretations, we can remember ourselves. Today's interpretive communities of practice (see also Denzin & Lincoln, 2018) are shaped by different interpretative pasts, and thus, the futures they create are diversified but still moving toward transformation and a more just world.

References

Anker, E., & Felski, R. (Eds.). (2017). *Critique and postcritique*. Duke University Press.

Bachelard, G. (2014). *The poetics of space* (new ed.). Penguin Classics.

Denzin, N., & Lincoln, Y. (2018). Introduction: The discipline and practice of qualitative research. In N. Denzin & Y. Lincoln (Eds.), *The SAGE handbook of qualitative research* (Vol. 5, pp. 1–26). SAGE.

Grant, A. (2014). Troubling 'lived experience': A post-structural critique of mental health nursing qualitative research assumptions. *Journal of Psychiatric and Mental Health Nursing, 21*(6), 544–549. https://doi.org/10.1111/jpm.12113.

Grbich, C. (2013). *Qualitative data analysis*. SAGE.

Koro-Ljungberg, M. (2016). *Reconceptualizing qualitative research: Methodologies without methodology.* SAGE.

Lincoln, Y. S., Lynham, S. A., & Guba, E. G. (2018). Paradigmatic controversies, contradictions, and emerging confluences, revisited. In N. K. Denzin & Y. S. Lincoln (Eds.), *The SAGE handbook of qualitative research* (pp. 108–150). SAGE.

Povinelli, E. (2016). *Geontologies: A requiem to late liberalism*. Duke University Press.

Saldaña, J. (2015). *Thinking qualitatively: Methods of mind*. SAGE.

Serres, M. (2008). *The five senses: A philosophy of mingled bodies*. Continuum.

Somekh, B., & Schwandt, T. (Eds.). (2007). *Knowledge production: Research work in interesting times*. Routledge.

Springgay, S., & Truman, S. E. (2017). On the need for methods beyond proceduralism: Speculative middles, (in)tensions, and response-ability in research. *Qualitative Inquiry*, 1–12. https://doi.org/10.1177/1077800417704464.

Tuck, E., & McKenzie, M. (2015). Relational validity and the 'where' of inquiry: Place and land in qualitative research. *Qualitative Inquiry*, 1–6. https://doi.org/10.1177/1077800414563809.

Vagle, M. (2014). *Crafting phenomenological research*. Left Coast Press.

Van Manen, M. (2014). *Phenomenology of practice.* Left Coast Press.

Willig, C. (2014). Interpretation and analysis. In Uwe Flick (Ed.), *The SAGE handbook of qualitative data analysis* (pp. 136–149). SAGE.

20 Voice

Abigail Hackett and Michael Gallagher

Interdisciplinary Development

In qualitative research, 'voice' usually refers to a concern to foreground the experiences and viewpoints of research participants. Participatory and co-produced research involves efforts to attend to the 'voices' of participants themselves, to avoid the risk of the researcher 'speaking for others' (Alcoff, 1991, p. 5), and in so doing, misrepresenting, misinterpreting, homogenising or silencing lives and perspectives different from their own. As such, voice is driven by commitments that are as follows:

> Epistemological: participants are seen as a valuable source of knowledge about their lives. Interpreting the meanings of what participants say is seen as a way to increase the validity of qualitative research.
>
> Political: participants are seen as having a democratic right to speak and be heard.
>
> Ethical: creating space for participants to 'speak their truth' is seen as a practical way to enact ethical principles such as respect, justice and empowerment.

Voice has been particularly prominent in subdisciplines concerned with marginalised groups, such as disability studies, social work research and childhood studies. In these fields, qualitative research on voice links up with wider activities outside academia, including community education, advocacy, activism and policy development. In childhood studies, for example, the UN Convention on the Rights of the Child has become a touchstone for considerations relating to children's voices across research, policy and practice. Article 12 stipulates that children have the right to express their views on matters affecting them. In recent decades, a vast array of initiatives, activities and research projects have formed around that impetus. There are differences of emphasis between different fields in the terminology relating to voice, although there is also considerable overlap. In disability studies, voice is often discussed in relation to inclusion. In childhood studies, the dominant concept is that of

DOI: 10.4324/9781003585046-21

participation (e.g. Hill et al., 2004; Moran-Ellis & Tisdall, 2019; Correia et al., 2021). In health services research and social work research, voice is frequently framed in terms of service user involvement (e.g. McLaughlin, 2010). In education, there is a field of research and practice on pupil voice and student voice (e.g. Cook-Sather, 2020). In early years, emphasis has been placed on listening to children's voices through multisensory methods in ways that align with and borrow from early years pedagogy (e.g. Clark et al., 2005; Wall & Robinson, 2022).

There have also been critiques of the notion of 'voice' as a solution to addressing power imbalances, making research ethical and hearing the experiences of others (Jackson & Mazzei, 2009; Marker, 2009; Smith, 1999). 'Voice' is often used to shore up the apparatus of interpretation in qualitative research, producing a sense of telling 'authentic' stories about people's lives (e.g. Schratz, 2020), but at the cost of sidelining aspects of voice that do not fit neatly into that apparatus. As Smith (1999) argues in relation to decolonising methodology, we cannot talk about reclaiming voice without also 'reordering those ways of knowing which were submerged, hidden or driven underground' (p. 79). What is meant by 'voice' often rests on assumptions that privilege 'authentic', coherent self-expression in language (Komulainen, 2007). These assumptions simultaneously marginalise voices that do not conform to dominant frameworks of communication (Ashby, 2011). Thus, voice within qualitative research has paradoxically been positioned as both the route to empowering research participants and a way in which participants can be exploited or excluded. In practice, the potential of voice often becomes reduced to the forms of knowledge that are most easily consumed by the machinery of qualitative research.

Voice, Relation and Power

> I turn to the other, with a voice shaped by *this other one*: I speak in order to locate myself near you.
>
> (LaBelle, 2014, p. 3; emphasis in the original)

That voice is relational rather than individual, provisionally expressing contingent meanings rather than firmly stating fixed positions, was a key argument in Spivak's (1987) influential contribution to postcolonial theory, *'Can the subaltern speak?'* Understanding voice as relational underscores the significance of an exchange or dialogue between two or more individuals. What one is able to say, and how it is heard or interpreted by the listener, is imbricated in relations of power and privilege that extend far beyond the individuals involved (Alcoff, 1991). From the speakers' perspective, what is and is not possible to express is fluid and shaped by the situation. Similarly, the researcher can only hear and interpret voices from their own situated position, resulting in an unavoidable process of researchers 'shaping participants' "exact words" through the unequal power relationships present and by our own research agendas and timelines' (Mazzei & Jackson, 2012, p. 746). Dialogue is never a neutral encounter, but

> deeply entwined with assumptions about communicative norms and rationality that privilege those from particular classed, gendered and racialised positions and invalidate the speech of others.
>
> (Mayes, 2019, p. 1197)

How these relational conditions of dialogue play out is at the heart of attempts to rethink voice in qualitative research and debates about how such conceptualisations can be put into practice methodologically.

The Problems With the Misuse of 'Voice'

Under-theorised within qualitative research, voice risks becoming a metaphor for participants' rights and authentic truths (Gallagher, 2020). This is problematic for several reasons. Firstly, the notion of authenticity is underpinned by conceptualisations of truth as universal and discoverable, rather than an understanding of knowledge as 'situated, split, emerging, tenuous and "stitched together imperfectly"' (Haraway, 1991, p. 193). If truth is fixed and universal, once it has been 'discovered' by methodologies that aim to interpret the 'authentic' meaning of voices, it ceases to matter who speaks this truth (or reports these findings). In this way, the line between speaking 'for' others and speaking 'about' others becomes increasingly blurred (Alcoff, 1991). The dangers of the extractive nature of 'voice' were eloquently outlined by bell hooks (1990), who described the appropriation of Black women's voices by white women, who then remade them in their own image:

> [There is] no need to hear your voice, when I can talk about you better than you can speak about yourself. . . . Re-writing you, I write myself anew. I am still author, authority. I am still [the] colonizer, the speaking subject, and you are now at the centre of my talk.
>
> (hooks, 1990, p. 152)

Drawing on hooks (1990) and Spivak (1987), Alcoff (1991, p. 6) argues that 'where one speaks from affects the meaning and truth of what one says' and that many privileged locations are 'discursively dangerous' in that they carry an increased risk of misrepresenting, taking up space to the detriment of others and doing more harm than good.

Voice Beyond Rational Signification

Often, methodologies designed to foreground the 'voices' of participants work most smoothly when those 'voices' mostly wish to express rational and abstractable knowledge in language familiar to the researcher (Jackson & Mazzei, 2009; Mazzei & Jackson, 2012). Often, 'in the research event, speech may still remain within the bounds of what the researcher can hear' (Mayes, 2019, p. 1197). Across fields including childhood studies (Spyrou, 2016; Warming, 2011), youth studies (Kraftl, 2013) and health (Spencer et al., 2020), scholars have raised concerns that a methodological interest in voice can 'struggle to accommodate desires or perspectives that are uncertain, unclear, partially formed, or which fall outside of binary choices' (Gallagher, 2020, p. 452).

The conflation of voice with authenticity relies on the idea of a bounded and singular speaking subject, who 'knows who she is, says what she means and means what she says' (MacLure, 2009, p. 104). As Mazzei and Jackson (2012) point out, when the focus remains on voice as a route to authentic subject positions, qualitative research will always prefer a voice that is 'present, stable, and self-reflective.' This results in a kind of hierarchy of 'voice' in which some perspectives or modes of expression become over-represented at the expense of others (e.g. see Spencer et al., 2020). In turn, methodological effort can turn to techniques most likely to 'produce' such voices; lack of clarity, seeming lack of 'truthfulness' or double meanings are viewed as methodological glitches to be erased in the process of producing valid interpretations (MacLure, 2013). Certain kinds of responses are more 'useful' to a researcher, for example, where they offer up readily interpretable meanings, address research questions directly or support the researcher's own held beliefs and values (Millei & Rautio, 2017). As Yoon and Templeton (2019) point out, when certain responses are valued over others in classroom contexts, it becomes difficult to really listen to children, to 'hear them out' (p. 55).

Researchers are increasingly interested in asking: what would it mean to work with a conceptualisation of voice that deliberately seeks to extend beyond what is comfortable, interpretable or understandable for the researcher concerned?

Voice, the Body and Being Human

The problem of how to consider voice beyond a hierarchy of signification returns us to the narrow and vaguely defined notion of 'voice' itself outlined earlier. Via the dominating position of white Western middle-class language practices (Avineri et al., 2015), 'certain modalities of speech . . . have been ignored as unintelligible' (Phipps, 2019,

p. 26). Notable within this hierarchisation of languages and forms of speech is an enduring kind of 'disgust' for the language practices of those less powerful, including a frequent tendency to compare them to animal noises (i.e. less than human) (Maclure, 2009). Additionally, forms of meaning-making beyond language are frequently overlooked (Flewitt, 2005) or else treated as a bridge to eliciting verbal responses (such as asking participants to draw their responses and then verbally describe the drawing).

Materialist accounts of language emphasise the way in which language issues from the body, from tongues, mouths and vocal cords. MacLure (2013) draws on Deleuze to point out that language is a bodily act and that there is always something within language that exceeds representation or meaning and resists interpretation. In order to move beyond these issues of which meanings are considered most meaningful, most useful and easiest to interpret, one way forward might be to engage in what MacLure (2013) has called 'the materiality of language' (p. 658).

In qualitative research, words tend to become separated from bodies (Lecercle, 2002) so that material realities become obscured behind representational linguistic systems. Often within qualitative research design, data emphasise what people say (e.g., interviews and focus groups); however, frequently 'the bodily entanglements of language [are considered] troublesome or trivial' (Maclure, 2013, p. 664). For example, interviews are neatly transcribed, thematic codes are identified, and key messages are collated in order to identify what participants *meant* when they spoke. Lecercle's (2002) work refers to bloodless angels within qualitative research; that is speakers with words but no bodies. How might we avoid creating 'boring bloodless angels' (MacLure, 2013, p. 665) within qualitative research concerned with participant's voice? MacLure suggests a focus on 'quasi-linguistic stuff' such as 'all the tears, sneers, sighs, silences, sniffs, laughter, snot, twitches or coughs that are part of utterances' (p. 664), precisely because they present a challenge to dominant modes of what it means to know. In this way, they disrupt the notion of voice as an 'authentic' representation of how people 'truly' think. A focus on 'voice' as bodily and more-than-representational unsettles the traditional approach to qualitative research that aims to interpret what participants 'really' mean and then diagnose a lack (in either the participants or the research design) when this task is difficult to achieve.

Unravelling Voice: A More Vulnerable and Less Certain Research Practice

This piece has explored the implications for qualitative research of more critically interrogating voice and how it has been theorised. Through such interrogation, the role of voice as an authentic representation of how individuals truly feel and think unravels, leaving the qualitative researcher in a more precarious and less certain position.

In her account of multilingualism, Phipps (2019) describes situations in which individuals do not share the same language as a space of partial knowing. Whilst acknowledging the struggle this kind of situation involves, Phipps makes a compelling argument for the shifts in power and relationships that can occur as

> language . . . swirls and forms and falls around what is already known, and the desire to understand.
>
> (Phipps, 2019, p. 43)

This kind of unravelling creates a vulnerable space that can encourage the researcher to move from a stance of authority to a position of puzzling, wondering and being willing to listen to the voices of others in ways that might not have been anticipated in advance (Badwan, 2022). As such, it offers a hopeful and generative position to think from and a potential starting point for more radically open-ended forms of interpretation in qualitative research.

References

Alcoff, L. (1991). The problem of speaking for others. *Cultural Critique*, 20, 5–32. https://doi.org/10.2307/1354221.

Ashby, C. (2011). Whose "voice" is it anyway?: Giving voice and qualitative research involving individuals that type to communicate. *Disability Studies Quarterly*, 31(4), 1–21.

Avineri, N., Johnson, E., Heath, S., McCarty, T., Ochs, E., Kremer-Sadlik, T., Blum, S., Zentella, A. C., Rosa, J., Flores, N., & Alim, H. S. (2015). Invited forum: Bridging the "language gap." *Journal of Linguistic Anthropology*, 25(1), 66–86.

Badwan, K. (2022) 'I'm afraid there are no easy fixes': Reflections on teaching intercultural communication through embracing vulnerability. In P. Holmes & J. Corbett (Eds.), *Critical Intercultural Pedagogy for Difficult Times: Conflict, Crisis and Creativity*. Routledge.

Clark, A., Kjorholt, A. T., & Moss, P. (Eds.). (2005). *Beyond Listening: Children's Perspectives on Early Childhood Services*. The Policy Press.

Cook-Sather, A. (2020). Student voice across contexts: Fostering student agency in today's schools. *Theory Into Practice*, 59(2), 182–191.

Correia, N., Aguiar, C., & Amaro, F. (2021). Children's participation in early childhood education: A theoretical overview. *Contemporary Issues in Early Childhood*. https://doi.org/10.1177/1463949120981789.

Flewitt, R. (2005). Is every child's voice heard? Researching the different ways 3-year-old children communicate and make meaning at home and in a pre-school playgroup. *Early Years*, 25(3), 207–222.

Gallagher, M. (2020). Voice audio methods. *Qualitative Research*, 20(4), 449–464.
Haraway, D. (1991). *Simians, Cyborgs, and Women*. Routledge.
Hill, M., Davis, J., Prout, A., & Tisdall, K. (2004) Moving the participation agenda forward. *Children & Society*, 18, 77–96.
hooks, b. (1990). *Yearning: Race, Gender, and Cultural Politics*. South End Press.
Jackson, A. Y., & Mazzei, L. A. (Eds.). (2009). *Voice in Qualitative Inquiry: Challenging Conventional, Interpretative and Critical Conceptions in Qualitative Research*. Routledge.
Komulainen, S. (2007). The ambiguity of the child's 'voice' in social research. *Childhood*, 14(1), 11–28.
Kraftl, P. (2013). Beyond 'voice', beyond 'agency', beyond 'politics'? Hybrid childhoods and some critical reflections on children's emotional geographies. *Emotion, Space and Society*, 9, 13–23.
LaBelle, B. (2014). *Lexicon of the Mouth: Poetics and Politics of Voice and the Oral Imaginary*. Bloomsbury.
Lecercle, J.-J. (2002). *Deleuze and Language*. Palgrave Macmillan.
MacLure, M. (2009). Broken voices, dirty words: on the productive insufficiency of voice. In A. Jackson & L. Mazzei (Eds.), *Voice in Qualitative Inquiry: Challenging Conventional, Interpretive, and Critical Conceptions in Qualitative Research* (pp. 97–113). Routledge.
MacLure, M. (2013). Researching without representation? Language and materiality in post-qualitative methodology. *International Journal of Qualitative Studies in Education*, 26(6), 658–667.
Marker, M. (2009). Indigenous voice, community, and epistemic violence: The ethnographer's 'interests' and what 'interests' the ethnographer. In A. Jackson & L. Mazzei (Eds.), *Voice in Qualitative Inquiry: Challenging Conventional, Interpretive, and Critical Conceptions in Qualitative Research* (pp. 27–43). Routledge.
Mayes, E. (2019). The mis/uses of 'voice' in (post)qualitative research with children and young people: Histories, politics and ethics. *International Journal of Qualitative Studies in Education*, 32(10), 1191–1209.
Mazzei, L. A., & Jackson, A. Y. (2012). Complicating voice in a refusal to 'let participants speak for themselves'. *Qualitative Inquiry*, 18(9), 745–751.
McLaughlin, H. (2010). Keeping service user involvement in research honest. *The British Journal of Social Work*, 40(5), 1591–1608.
Millei, Z., & Rautio, P. (2017) 'Overspills' of research with children: an argument for slow research. *Children's Geographies*, 15(4), 466–477.
Moran-Ellis, J., & Tisdall, E. K. M. (2019). The relevance of 'competence' for enhancing or limiting children's participation: Unpicking conceptual confusion. *Global Studies of Childhood*, 9(3), 212–223.
Phipps, A. (2019). *Decolonising Multilingualism: Struggles to Decreate*. Multilingual Matters.
Schratz, M. (2020). Voices in educational research: An introduction. In M. Schatz (Ed.), *Qualitative Voices in Educational Research* (pp. 1–6). Routledge.

Smith, L. T. (1999). *Decolonizing Methodologies: Research and Indigenous Peoples*. Zed Books.
Spencer, G., Fairbrother, H., & Thompson, J. (2020). Privileges of power: Authenticity, representation and the "problem" of children's voices in qualitative health research. *International Journal of Qualitative Methods*, 19, 1–8.
Spivak, G. C. (1987). Can the subaltern speak? In C. Nelson & L. Grossberg (Eds.), *Marxism and Interpretation of Culture* (pp. 271–313). University of Illinois Pres.
Spyrou, S. (2016). Researching children's silences: Exploring the fullness of voice in childhood research. *Childhood*, 23(1), 7–21.
Wall, K., & Robinson, C. (2022). Look who's talking: Eliciting the voice of children from birth to seven. *European Early Childhood Education Research Journal*. Online first.
Warming, H. (2011) Getting under their skins? Accessing young children's perspectives through ethnographic fieldwork. *Childhood*, 18(1), 39–53.
Yoon, H., & Templeton, T. (2019). The practice of listening to children: The challenges of hearing children out in an adult-regulated world. *Harvard Educational Review*, 89(1), 55–84.

21 Wriggling With Data

Teija Löytönen and Henrika Ylirisku

Data, the central concept and element in research practices, have become problematic. Data in the research fields of arts, education, and art education, where this entry is grounded, often lean on conventional humanistic qualitative inquiry, where data still denote quite narrowly empirical materials, samples, and extracts collected as evidence "from the field" via formal and systematic procedures (e.g. interviews and observations), to be then analyzed and interpreted. The engagements with post-structural and post-humanist theories, relational onto-epistemologies, and post-qualitative approaches have both challenged and motivated scholars to (re)consider the doing of the research as "an enactment of knowing-in-being that emerges in the event of doing research itself" (Taylor, 2016, p. 18) and move beyond predefined notions of data and procedural methods (Koro-Ljungberg et al., 2017; St. Pierre et al., 2016; St. Pierre, 2021; Springgay & Truman, 2018). Instead of abandoning data altogether, we, among other scholars, have wanted to experiment in several research processes with how data could transform toward "relational data, more-than-human data, experimental data, performative and embodied lifedata, data that are so entangled with life that they become multiple" (Koro et al., 2022). Data, then, are not stable or prefixed by "an objectivist epistemology and a realist ontology" (Rautio, 2021, p. 228) but "theoretically formed and ontologically situated" (Koro et al., 2022), and emerge differently in each singular research process, such as in the following examples.

The first example is doctoral research on environmental art education (Ylirisku, 2021). The aim of the research was to reconceptualize the theoretical groundings of environmental art education through posthumanist theories (e.g. Barad, 2007; Braidotti, 2013; Haraway, 2008; Malone, 2015; Neimanis, 2017), especially expanding the critique of the human exceptionalist/anthropocentric premises of the existing frames.

In addition to the theoretical reorientation, the research was carried out through an experiment called "Becoming-with the forest". In this

DOI: 10.4324/9781003585046-22

experiment, the post-humanist theories and concepts were put to work by taking them concretely to the forest, which agitated new thinking through embodied engagement. Regular orienteering (sports) practice in local forests functioned as a propositional catalyst. The practice was not called a method because "propositions are different from research methods or a research design in that they are speculative and event oriented" as Springgay and Truman (2018, p. 83) point out. One could say that the orienteering practice functioned as a springboard for getting physically tangled in relations and thinking with nonhuman others in/with the forest, becoming-with (Haraway, 2008) multispecies and material encounters.

The notions by Springgay and Truman (2018) on research as inhabiting "speculative middles" were particularly inspiring for the experiment. They highlight that research becomes a site of experimentation where things start to unfold from the process itself. In the "speculative middles", the research methods cease to serve for collecting data to be interpreted, analyzed, explained, reflected, and reported. Instead, methods and data emerge and unfold in entangled relations of being-thinking-moving in the middle (Springgay & Truman, 2018). Likewise, essential in the experiment was the problematizing of the researcher subject as a separate, exceptional human individual. The previous experiences in embodied and movement-based artistic (research) practices allowed the researcher to acknowledge and pay attention to the more-than-verbal dimensions of the human-forest entanglements: affective intensities, hapticality, touch, smell, and sound.

What was important was to stay open to whatever might come up (bites of insects, touches of vegetal bodies, sense of cold rain) and call for attention both by leaning on systematic practices and by engaging with unforeseen encounters/invitations/provocations (see Figure 21.1). These served as data for participation in the world through and with the materials on offer (Rautio, 2021, p. 228). Physical orienteering practice, writing journal entries, reading post-humanist theory, taking documentary photographs, and studying further certain encounters in the forest through making photos with a photographer, turning then into visual-textual stories, all of these becoming data.

> iiiiiiiiiiiiiiiiiiiiilllllllllliiiiiiiiilllllliiiiiiiiiiiiiiiilllllllliiiiiiiiiiiiiiiiiiiiii
> Hear them too?
> Even their whine is forbidding.
> . . .
> (I) do not want to share (my) body with you!
> You question (my) bodily integrity TOO MUCH!
>
> (Ylirisku, 2021, p. 140)

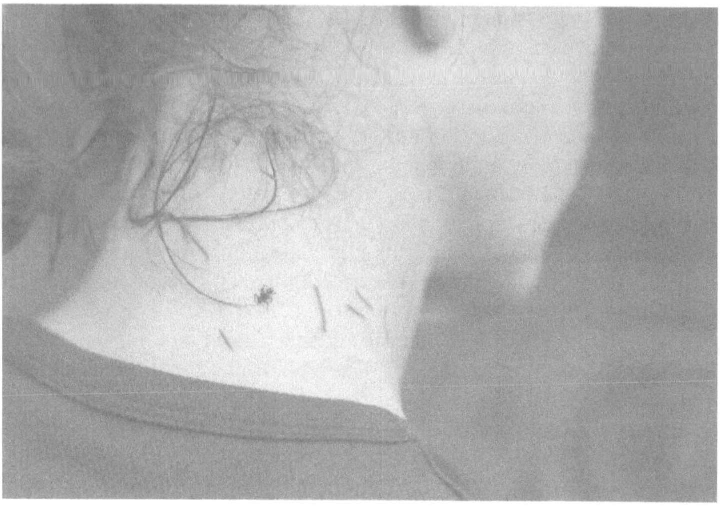

Figure 21.1 Ylirisku, Henrika (2019). *Becoming-with insects*
Source: (photo: Ylirisku, Pasi; Ylirisku, 2021)

Another example is the study on educational development within higher arts education, which focused on exploring differences in university pedagogy in the arts (Löytönen, 2017). Based on a three-year pedagogical course, the aim was to question fixed pedagogical concepts and harmonization practices. The experimentation was framed by the philosophy of Gilles Deleuze (1994) and especially his concept of difference. Deleuze invited us to move beyond the simplifications of dualisms, binaries, categorizations, or oppositions (e.g. deep/surface learning and foundational/applied disciplines) toward free differences and multiplicities: difference in the constant state of becoming or differentiation, "things" becoming different in themselves.

During the pedagogical course, the participants created images and written notes, poems, video clips, photographs, recorded sounds, "and . . . and . . . and" (Deleuze & Guattari, 1987, p. 25) on university pedagogy. These data were understood through the Deleuzian notion of event, which emerges when "[s]omething in the world forces us to think. This something is an object not of recognition but of a fundamental *encounter*" (Deleuze, 1994, p. 139, emphasis in the original). Thus, events, Deleuze (2004) writes, "are actualized within us, they wait for us and invite us in" (p. 169). What is needed is "to be attentive and open to surprise to recognize the invitation. And once invited in . . . to experiment and see where it takes us" (MacLure, 2013, p. 231; see also

Koro-Ljungberg, 2016; Koro-Ljungberg & MacLure, 2013). Data, then, do not have fixed or stable meanings to be described, explained, and interpreted. Instead, data are active and engaging, inviting scholars to think/explore/sense with.

The encounters with these data, the diverse creations, and the events invited (forced) the researcher to think (about) how to explore the differentiation or becoming in university pedagogy. Here, the notion by May (2005) proved helpful. He wrote that "a central feature of Deleuze's ontology of difference is that difference can be palpated but not brought into the categories of representation" (p. 95). Palpation then is thinking in the encounters with the creations, in this experiment, about differentiation in pedagogy. It was carried out through the speculative question of *what-if*. Following Manning and Massumi (2014), "it is never a question of formally working something out in advance" (p. 43). Instead, the *what-if* question is allied to the potential: it is tentative, open, and partial, drawing attention to connections, movement (flux), and the constant becoming of things (see also St. Pierre, 2013). Palpation enabled sense to emerge (Masny, 2016) toward multiplicities in university pedagogy and higher arts education.

> What if educational development becomes aesthetics – eternal beauty – focusing on
> pedagogical aesthetics?
> . . .
> What if we gave up the privileged position of the logic of rationality (pedagogy as only a cognitive or conceptual activity) for the "logic of sense" (Deleuze, 2004)?
> . . .
> What if pedagogy becomes a Möbius strip?
> (Löytönen, 2017, pp. 239–240)

The previous experiments allowed realizing that data in research processes are always already present, providing multiple, vivid, and strong materials to think-become-move with. Thus, data have shifted from "clearly defined evidence from the field" toward data as entangled in relations with theories, emerging through situated research events and researcher subjectivities. In diverse research processes, there might be some anticipated orientations on how to engage with the data, but at the same time, it is important to welcome serendipitous surprises and invitations that might mobilize/disturb/provoke new insights on thinking with the data. These in turn call for the researchers to attune to new sensibilities, openness, and readiness to respond to the messy invitations, events, and relations in and of the world. Researchers, thus, are challenged to reorient themselves toward future directions or speculations in interpretation and data in each specific research process.

References

Barad, K. (2007). *Meeting the universe halfway: Quantum physics and the entanglement of matter and meaning*. Durham, NC: Duke University Press.

Braidotti, R. (2013). *The posthuman*. Cambridge: Polity Press.

Deleuze, G. (1994). *Difference and repetition* (P. Patton, Trans.). New York: Columbia University Press.

Deleuze, G. (2004). *The logic of sense* (M. Lester, Trans.). London: Bloomsbury.

Deleuze, G., & Guattari, F. (1987). *A thousand plateaus: Capitalism and schizophrenia* (B. Massumi, Trans.). Minneapolis, MN: University of Minnesota Press.

Haraway, D. (2008). *When species meet*. Minneapolis, MN: University of Minnesota Press.

Koro, M., Fairchild, N., Benozzo, A., & Löytönen, T. (2022). Qualitative sampling and qualitative data. In R. Tierney, F. Rizvi, K. Ercikan, & G. Smith (Editors-in-Chief), *International encyclopaedia of education, fourth volume* (pp. 198–209). Oxford: Elsevier.

Koro-Ljungberg, M. (2016). *Reconceptualizing qualitative research: Methodologies without methodology*. Los Angeles, CA: Sage.

Koro-Ljungberg, M., Löytönen, T., & Tesar, M. (Eds.). (2017). *Disrupting data in qualitative inquiry: Entanglements with the post-critical and post-anthropocentric*. New York: Peter Lang.

Koro-Ljungberg, M., & MacLure, M. (2013). Provocations, re-un-visions, death and other possibilities of 'data.' *Cultural <–> Critical Methodologies*, 13(4), 219–222.

Löytönen, T. (2017). Educational development within higher arts education: An experimental move beyond fixed pedagogies. *International Journal for Academic Development*, 22(3), 231–244.

MacLure, M. (2013). The wonder of data. *Cultural Studies – Critical Methodologies*, 13(4), 228–232.

Malone, K. (2015). Posthumanist approaches to theorizing children's human-nature relations. In K. Nairn, P. Kraftl, & T. Skelton (Eds.), *Space, place and environment* (pp. 1–22). Singapore: Springer. https://doi.org/10.1007/978-981-4585-90-3.

Manning, E., & Massumi, B. (2014). *Thought in the act: Passages in the ecology of experience*. Minneapolis, MN: University of Minnesota Press.

Masny, D. (2016). Problematizing qualitative research: Reading a data assemblage with rhizoanalysis. *Qualitative Inquiry*, 22(8), 666–675.

May, T. (2005). *Gilles Deleuze: An introduction*. Cambridge: Cambridge University Press.

Neimanis, A. (2017). *Bodies of water: Posthuman feminist phenomenology*. New York: Bloomsbury Academic. https://doi.org/10.1353/mis.1987.0051.

Rautio, P. (2021). Post-qualitative inquiry: Four balancing acts in crafting alternative stories to live by. *Qualitative Inquiry*, 27(2), 228–230. https://doi.org/10.1177/1077800420933297.

Springgay, S., & Truman, S. E. (2018). *Walking methodologies in a more-than-human world: WalkingLab*. Oxford and New York: Routledge.
St. Pierre, E. A. (2013). The posts continue: Becoming. *International Journal of Qualitative Studies in Education*, 26(6), 646–657. https://doi.org/10.1080/09518398.2013.788754.
St. Pierre, E. A. (2021). Why post qualitative inquiry? *Qualitative Inquiry*, 27(2), 163–166. https://doi.org/10.1177/1077800420933297.
St. Pierre, E. A., Jackson, A. Y., & Mazzei, L. A. (2016). New empiricisms and new materialisms: Conditions for new inquiry. *Cultural Studies ↔ Critical Methodologies*, 16(2), 99–110. https://doi.org/10.1177/1532708616638694.
Taylor, C. A. (2016). Edu-crafting a cacophonous ecology: Posthumanist research practices for education. In C. A. Taylor & C. Hughes (Eds.), *Posthuman research practices in education* (pp. 5–24). Hampshire: Palgrave Macmillan.
Ylirisku, H. (2021). *Reorienting environmental art education* (Doctoral Dissertation). Espoo, Finland: Aalto ARTS Books. Aalto University Publication Series Doctoral Dissertations 9/2021.

Index